Fatigue Thresholds

David Taylor
Department of Mechanical and Manufacturing Engineering, Trinity College, Dublin

Butterworths
London Boston Singapore Sydney Toronto Wellington

 PART OF REED INTERNATIONAL P.L.C.

British Library Cataloguing in Publication Data

Taylor, D. (David), *1956–*
    Fatigue thresholds.
    1. Materials. Fatigue
    I. Title
    520.1'123

ISBN 0-408-03921-3

Library of Congress Cataloging in Publication Data

Taylor, David.
    Fatigue thresholds.
    Includes bibliographical references.
    1. Materials—Fatigue.   2. Fracture mechanics
I. Title.
TA418.38.T38   1989   620.1'126   89-17411
ISBN 0-408-03921-3

Photoset by Butterworths Litho Preparation Department
Printed and bound in Great Britain by Courier International Ltd, Tiptree, Essex

DETAILS FOR BORROWER'S SLIPS:

Auth/Ed: TAYLOR

Title: Fatigue thresholds

Vol: Copy No:

Stamp indicate
returns wil'

Preface

This book is concerned with a particular fracture-mechanics parameter, ΔK_{th}, known as the 'threshold stress intensity range', or 'fatigue threshold'. It is also concerned, on a more general level, with high-cycle 'limits' to fatigue in materials, whether these are expressed in terms of the old standard fatigue limit, the LEFM parameters or other more recent, variants.

It is now accepted that long fatigue life in materials and components almost always involves 'living with cracks', that the threshold or limit condition is associated with the non-propagation of existing cracks or defects, even though these cracks may be microscopic in size and undetectable in components. Therefore, some form of fracture mechanics is required, some analysis involving stress and crack length as well as a variety of other parameters relating to the material, the environment and the loading type.

This book outlines the use of linear elastic fracture mechanics and other approaches to the description of fatigue crack propagation in the high-cycle regime. Early chapters review current theories on near-threshold behaviour and known mechanisms of slow crack advance. After discussion of methods of measurement and appropriate standards in this area, the effect of individual parameters (mechanical properties, microstructure, environment and load type) are treated in detailed chapters. Two chapters are devoted to the difficult areas of defect type, notches and short cracks. Finally, the applicability of ΔK_{th} as a threshold parameter is discussed, and various case studies are presented to illustrate its practical use.

It is hoped that the book will be of value in advanced courses in the areas of materials engineering, materials science and failure analysis, and also to academic researchers in these disciplines, but most of all the book is directed to practising engineers.

While attempting to give adequate coverage of theoretical aspects of the topic, and to substantiate all comments with reference to the literature, my main preoccupation in writing this book has been to put myself in the place of engineers working in the areas of design, materials selection and component lifing. Many engineers remain understandably sceptical about the practical value of fatigue crack propagation concepts over more traditional approaches to integrity assurance. For this reason I have attempted, wherever possible, to suggest simple, conservative design rules.

I have also attempted to identify certain areas of the subject, such as environmental effects and variable-amplitude loading, which must still be approached with caution, and for which more research work is encouraged.

I am always interested to hear of the experiences of other workers in the use (or misuse) of threshold concepts in design, materials selection or failure analysis, and I would be more than willing to exchange views and advice in such cases. It is often difficult to obtain adequate data on thresholds and near-threshold growth rates; I have collected some data in the process of writing this book, and a previous publication, which are available in a computer database; I would appreciate it if anyone who has new data in this area would communicate it to me. Finally, I would welcome any comments (favourable or otherwise) on the material presented below.

This book represents the work of one long summer, and I would like to thank all those who endured my company, or lack of it, and countered my lapses of enthusiasm with encouragement, especially Niamh.

David Taylor
Trinity College, Dublin
November 1988

Contents

1 Introduction

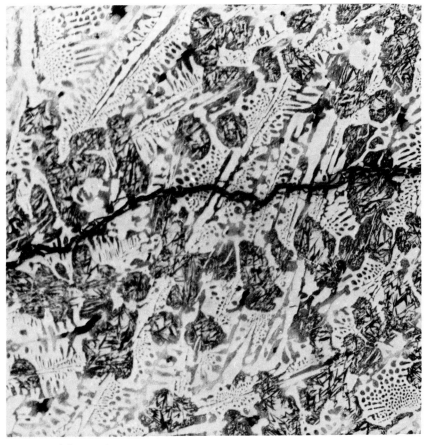

Fatigue crack in a cast iron

Limits to fatigue failure

If we wish to ensure that a component can withstand a large number of fatigue cycles, i.e. if we are designing in the high-cycle regime, then a parameter of paramount interest will be the 'limit' or 'threshold', which may be defined simply as the most severe cyclic load that can be applied without fatigue failure occurring. In practice, this definition may have to be modified to a requirement of non-failure within some large, finite number of cycles, but in many materials the behaviour at high numbers of cycles is such that a boundary occurs between 'fail' and 'non-fail' conditions at a specific value of stress or stress intensity.

The designer may then choose either to remain always below this parameter (possibly with the addition of a safety factor) or to use the material above its 'threshold', with the necessity of estimating a fatigue life in service. In either case the value of the 'threshold' parameter, and the behaviour of material when loaded just above it, will be paramount in the choice of material and will have consequences for the design of the component.

I place the word 'threshold' in inverted commas to distinguish it from the precise, fracture-mechanics parameter, usually written ΔK_{th} or ΔK_o. The greater part of this book will be devoted to the ΔK_{th} parameter, its definition, measurement and use, but it should be remembered that ΔK_{th} is one of a number of threshold-type parameters which may be used.

The most common threshold is of course the fatigue limit or endurance limit derived from an S/N curve. The fatigue limit has served us well for many years and may still be usefully applied, especially to well-understood materials and with the benefit of experience. However, it is generally recognized that S/N-based data is unsatisfactory in the choice of materials for many critical components because it is highly specimen-specific and, more fundamentally, because it is not mechanistically based. The specification of a total number of cycles to failure, or to the initiation of a macroscopic crack, takes no account of our understanding that fatigue is divided into the three processes of crack initiation, short-crack growth and long-crack growth. These processes involve different mechanisms and different constraints, and any one of them may dominate the high-cycle life.

Current thinking tends to the view that the crack initiation stage is by-passed in almost all materials, and certainly in most engineering components, and this leads us to the assumption that the threshold parameter should be a fracture mechanics parameter. The great watershed, however, comes between those situations which can be adequately described by linear elastic fracture mechanics (LEFM) and those cases for which LEFM is not valid. In some cases a related fracture-mechanics parameter such as ΔJ or cyclic crack-opening displacement may be more appropriate. In other cases, for instance fatigue from notches and fatigue of short cracks, it may be appropriate to return to a stress-based threshold, $\Delta\sigma_{th}$, while recognizing that this describes a crack propagation process.

Scope of this book

In defining the scope of this book, then, I have limited myself to defect tolerance, assuming a pre-existing crack or notch, and defining the threshold as the most appropriate parameter to describe the growth/no-growth condition under high-cycle fatigue loading. I attempt to show that, despite its limitations, the LEFM-based threshold, ΔK_{th}, gives a very satisfactory description of material behaviour in many cases; Chapters 2–7 are largely devoted to the theory and practice of this parameter. In Chapters 8 and 9 the use of K-based and stress-based parameters are compared for the cases of short cracks, notches and other types of defect. Finally, Chapter 10 presents some case studies and general comments on the applicability of threshold concepts to practical situations.

Linear elastic fracture mechanics

It is not proposed to derive the basic LEFM parameters as part of this work. For this the reader is referred to one of the many excellent texts on this subject such as Knott's 'Fundamentals of Fracture Mechanics'[1] or Parker's 'The Mechanics of Fracture and Fatigue'[2].

We shall begin with the stress intensity parameter, K, which in its simplest formulation is written:

$$K = \sigma \sqrt{(\pi a)} \tag{1.1}$$

where σ is the stress applied to a large, uniform plate containing an edge crack, length a or centre crack length $2a$. A K parameter can be defined for any geometry and opening mode, and uniquely characterizes the elastic stress singularity associated with the crack and the otherwise uniform stress field.

Stress intensity was originally formulated in order to define unstable crack propagation, causing brittle fracture, at the value of K_c. In 1961 Paris proposed[3] that its use could be extended to describe fatigue crack propagation in terms of the peak-to-peak range of K in the fatigue cycle, ΔK. Soon after, it was noted that the majority of the growth-rate plot conformed to an equation of the form:

$$\frac{da}{dN} = A(\Delta K)^m \tag{1.2}$$

where da/dN is the crack growth rate per cycle, A and m being constants. This simple empirical fit, giving as it does a straight line on a log/log plot, has been applied with remarkable success over the whole range of metallic materials, and in some cases to non-metallic materials.

It is interesting to note that Paris's original paper was rejected by several leading journals on the grounds of its lack of a mechanistic basis, specifically because it was applying elastic mechanics to a process of crack advance which is dominated by near-crack-tip plasticity. To this day there is still no sound theoretical basis for the Paris equation, which must be judged on its practical success as an empirical characterization. This is a point which should be kept in mind when the validity of LEFM parameters is discussed in the following chapters; although theoretical objections may be raised, the only sound test is the practical one.

The LEFM threshold value

Figure 1.1 shows one of the original data plots made by Paris, which, using data taken from existing pubications, convinced him of the usefulness of the K parameter. Even on this plot there is a clear indication of a threshold K value, below which the crack growth rate is negligable.

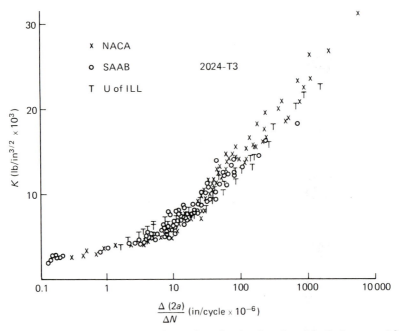

Figure 1.1 Some early crack growth-rate data, showing the value of the Paris approach [3]

In fact, fatigue crack propagation data are found to deviate asymptotically at both high and low ΔK values, producing the familiar sigmoidal curve, as shown in Figure 1.2. The higher asymptote is associated with rapid failure as the K_c value is approached. The lower asymptote is the threshold stress intensity range, denoted ΔK_{th} or ΔK_o, which is the principal character in this book. The line begins to curve towards the low asymptote at propagation rates in the region of 10^{-6} mm/cycle, becoming almost parallel to the da/dN axis below 10^{-7} mm/cycle. The detailed definition of the threshold ΔK_{th} will be discussed in Chapter 4 on methods of measurement. It is clear, however, that a value of ΔK can be defined below which propagation effectively ceases, and that for values of ΔK close to ΔK_{th} the crack propagation rate is a very strong function of ΔK (the effective Paris exponent being as high as 8–10 in many cases).

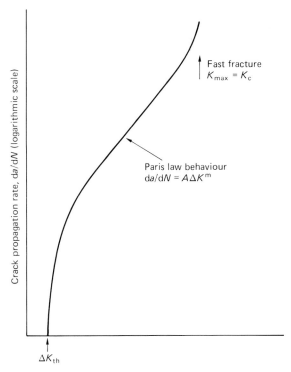

Crack propagation rate, da/dN (logarithmic scale)

Fast fracture
$K_{max} = K_c$

Paris law behaviour
$da/dN = A\Delta K^m$

ΔK_{th}

Figure 1.2 Schematic variation of crack growth rate with applied stress itensity range

It is convenient at this stage to define a number of parameters which will be used frequently in the following chapters:

K LEFM stress intensity;
K_{max} maximum value of K in a regular fatigue cycle;
K_{min} minimum value of K in a regular fatigue cycle;
ΔK peak-to-peak range of K, $\Delta K = K_{max} - K_{min}$;
K_{mean} mean value of K in the fatigue cycle;
R load ratio, K_{min}/K_{max}, also equal to $\sigma_{min}/\sigma_{max}$;
σ applied stress
$\sigma_{max}, \sigma_{min}, \sigma_{mean}, \Delta\sigma$ fatigue stress parameters as K_{max}, K_{min}, etc;
a crack length;
da/dN crack growth rate per cycle.

The values of ΔK_{th}, and of da/dN in the near-threshold region, effectively control high-cycle fatigue behaviour in most practical situations. This is because, in the first place, most real components and structures contain defects from which cracks may initiate relatively easily, making crack propagation the life-controlling process, and, second, because the nature of the growth rate dependence (Figure 1.2) implies that a crack will grow relatively slowly in the initial stages, accelerating rapidly as its length increases.

Now if the fatigue life is to exceed, say, 10^7 cycles, which is a common requirement for load-bearing components, the initial growth rate must lie within the near-threshold region, i.e. below the straight-line portion of the curve in Figure 1.2. Thus the behaviour of high-cycle components is critically dependant on the details of this asymptotic region.

Unfortunately, a feature of the near-threshold region is that it is strongly dependent on a number of variables which are not so influential in the straight-line region. These include the material's mechanical properties and microstructure, loading variables such as the R ratio, and environmental variables such as corrosion and temperature. In some ways this is a good thing, because it allows the possibility of alloy development to produce materials with high threshold values and low growth rates in the near-threshold region. However, the dependence on external variables implies a multi-parameter problem, requiring either detailed mechanistic understanding or extensive relevant data collection.

An added complication is that many material variables, such as grain size, have a profoundly different effect on ΔK_{th} from their effect on the fatigue limit as defined by S/N curves. In simple terms, crack 'initiation' and crack 'propagation' are controlled in different ways (though, as Chapter 8 will show, 'initiation' is often effectively controlled by the propagation of very short cracks).

This, then, defines the difficulties and challenges which face the application of threshold concepts. On the positive side, the benefits to be gained from an understanding of threshold behaviour are considerable. At present, high-cycle fatigue design is all too often based on highly conservative approaches, using fatigue limit data derived from S/N curves or, even more simply, from ultimate strength (e.g. $UTS/2$). Conservatism results in excessively large load-bearing areas, expensive alloys or too-frequent inspection periods.

The use of fracture-mechanics concepts offers, at least in principle, an exact calculation of the number of cycles required for a crack to grow from a given initial length to a definable dangerous length. In practice, the calculation is unlikely to be exact owing to uncertainty in various parameters such as the size of the initial defect and the value of the applied stress. The real advantage of the approach is that it is truly mechanistic, being linked to the physical process of crack growth.

At present, fracture-mechanics concepts, including ΔK_{th}, are used quite commonly in failure analysis, along with any other methods of analysis which may be able to explain the observed failure. Here it may be possible to be exact, if the initiating defect is measurable, and so an LEFM calculation will decide whether the observed defect could have propagated to failure under the known stress conditions.

The use of ΔK_{th} and other LEFM parameters in component design is evolving much more slowly. This is understandable, especially in the more traditional industries. At the time of writing, however, many aerospace industries now accept LEFM approaches as routine, especially in conjunction with NDE procedures. For example, some cracks in airframes and non-critical engine components can now be detected, defined as being below a certain critical length, and returned to service.

In the near future it is anticipated that similar philosophies will be used in, for example, the automotive and energy industries, especially given the importance of remnant life estimation in the latter. At present it is advised that the approach described in this book be used in *conjunction* with existing methodologies, as it is only by this test that its rightful areas of application will be identified.

Other reading

The reader is directed towards a number of comprehensive conference proceedings from recent years, which provide a full account of the development of this subject academically and also, in some cases, address the problems of practical application. These include the triennial series of conferences on 'Fatigue and Fatigue Thresholds'[4–6], the proceedings of the AIME conference *'Fatigue Crack Growth Threshold Concepts'*[7] and the EGF conference *'The Behaviour of Short Fatigue Cracks'*[8].

For a collection of data on threshold values and near-threshold growth rates the author is glad to recommend *'A Compendium of Fatigue Thresholds and Growth Rates'*[9] which contains over 1200 data sets covering most classes of metallic materials, loading conditions and environments.

References

1. Knott, J. F. (1973) *Fundamentals of Fracture Mechanics*, Butterworths, London
2. Parker, A. P. (1981) *The Mechanics of Fracture and Fatigue*, E. & F. N. Spon, London
3. Paris, P. C., Gomez, M. and Anderson, W. E. (1961) *Trends in Engineering 13,* University of Washington, Washington DC
4. *Fatigue Thresholds, Proceedings of the First International Conference on Fatigue Thresholds* (1982), EMAS, Warley, UK
5. *Fatigue 84, Proceedings of the Second International Conference on Fatigue Thresholds* (1985), EMAS, Warley, UK
6. *Fatigue 87, Proceedings of the Third International Conference on Fatigue Thresholds* (1988), EMAS, Warley, UK
7. Suresh, S. and Davidson, D. (eds) (1984) *Fatigue Crack Growth Threshold Concepts*, Proceedings of AIME Conference, The Metallurgical Society, Warrendale, PA
8. *The Behaviour of Short Fatigue Cracks, Proceedings of the EGF Conference* (1986), MEP, Bury St Edmunds, UK
9. Taylor, D. (1985) *A Compendium of Fatigue Thresholds and Growth Rates,* EMAS, Warley, UK

2 Theories and mechanisms

Fatigue fracture surface showing 'structure sensitive' near-threshold crack growth

Introduction

This chapter reviews the various theoretical models which have been used to describe near-threshold crack growth. It also discusses mechanisms of crack advance which have been observed, inferred from observations, or proposed for use in theoretical models.

Although it is hoped that this chapter will be of general interest, it is not necessary to read it in order to understand the rest of the book. Some readers may prefer to miss out this and the following chapter if their interests lie only in the experimental measurement and use of threshold data.

Crack closure

The phenomenon of crack closure is now so widely discussed as a mechanism in near-threshold crack propagation that it merits a chapter to itself. Therefore, the subject of closure, in both its theoretical and practical aspects, is discussed in Chapter 3. It should be noted that most of the theories discussed below would require modification in order to allow for the effects of closure.

Background to theoretical developments

A number of remarks may be made which are appropriate to all theoretical models in this subject area. It is clear that, despite over 20 years of effort in this direction, no single theoretical approach has emerged which describes threshold behaviour comprehensively. This, however, is a problem which is shared with all but the simplest structure/property relationships in materials science, and is possibly exaggerated by the strongly microstructural nature of the phenomenon in this case, of which more will be said below.

The advantages to be gained from a good predictive model are considerable. Experimental threshold measurement is expensive and time-consuming; any method which would allow us to interpolate or extrapolate from existing data within a given alloy system would be greatly appreciated. At present there are a few classes of materials within the alloy steels and aluminium alloys for which the body of existing data is large enough to allow the user to make a reasonable prediction of the behaviour of a given alloy in a given thermomechanical condition; for most systems, testing remains the only possible approach.

A related problem is the poor quality of much published data in this area, especially data obtained some years ago. Near-threshold data can be highly sensitive to the test method used (see Chapter 4), and until national Standards, which are being developed, become generally accepted, the problem of reliability of data will remain. This naturally also creates a stumbling block in the development of theoretical models, since most workers would look to the general body of published data to test the predictive capacity of their theoretical models.

Finally, near-threshold behaviour is very much a multi-parameter problem, involving a large number of microstructural, mechanical and environmental variables which often act synergistically.

Microstructural nature of the problem

A clear distinction may be drawn between the near-threshold portion of the growth-rate curve and the straight-line, 'Paris' region: in the near-threshold region changes to the microstructure of the material will generally produce strong effects, whereas in the Paris region this is generally not the case.

The reason for this difference may be stated simply as follows: in the near-threshold region the scale of deformation, as represented by the plastic zone size and the crack-tip opening displacement, is of the same order of magnitude as that of common microstructural features such as grain size and precipitate spacing. It has been demonstrated for many materials that a transition in fracture mode occurs, with fracture surfaces showing microstructurally dominated features under near-threshold conditions. Often the grain structure of the material is clearly visible on the fracture surface.

Therefore, it is a primary requirement of any theoretical model that it should incorporate microstructural features. However, some of the models described below have proceeded satisfactorily by first defining a threshold based on a homogeneous continuum and then refining this lower-bound prediction upwards by introducing microstructural concepts.

Having said this, the use of microstructural concepts involves some difficulties of quantification. Not only is it more difficult to measure, for example, a grain size than a mechanical property such as yield strength, it is also more difficult to define such a parameter uniquely; in the case of grain size, for example, the use of a linear intercept technique is fairly well established, but many workers still fail to correct the linear intercept figure to allow for the three-dimensional nature of the grain, and this correction must in turn assume a grain shape. Consequently, linear intercept results from different sources may differ by up to a factor of 2. Other microstructural features such as lath width and packet size for martensitic/bainitic steels are very difficult to define and quantify accurately.

Some theoretical models

Generally speaking, the models described below can be said to concentrate on near-crack-tip conditions, postulating some mechanism of crack advance which depends on a minimum value of stress, strain, energy or some combination of these three. A microstructural barrier such as a grain boundary is commonly introduced. I will deal first with the energy-based approaches.

Another class of theories uses, instead, the shape of the growth-rate curve as a starting point, deriving the threshold by various forms of curve-fitting and data analysis. These will be dealt with at a later stage.

Energy-based models

A natural starting point is to use some modification of a Griffth-type energy balance to define the crack growth threshold. Under threshold conditions the crack is able to propagate forward by an infinitesimally small amount before it arrests. The reason for the crack arrest may be related to one of several events such as crack-tip blunting or dislocation rearrangement, but the controlling factor may relate to the energy conditions that apply *before* the arrest occurs.

Purushothaman and Tien[1] re-examined the Griffth-type energy balance, replacing the normal form for the plastic work term with an expression dependent on ΔK. This led to a value of ΔK, when the energy balance was satisfied, which was strongly dependent on the surface energy term and also on Young's modulus, E. The values of ΔK derived for various metals were of the same order of magnitude as common threshold values at high R, though, generally, the predictions were on the low side.

The present author[2] modified this approach, deriving the plastic work term from assumed stress and strain conditions within the plastic zone, as opposed to the macroscopic definition used by Purushothaman and Tien. The resultant threshold prediction was:

$$\Delta K_{th} = \left(\frac{2\gamma\pi E}{1 - v^2 - 0.47I} \right)^{\frac{1}{2}} \tag{2.1}$$

where γ is the surface energy for creation of new crack faces, v is Poisson's ratio and I is given by the integral:

$$I = \int_0^1 \frac{\sqrt{[x\,(1-x)]}}{(x+1)\left(x + \frac{\sigma_y}{E\epsilon_t} \right)} \, dx \tag{2.2}$$

where ϵ_t is the true fracture strain of the material.

Since the value of the integral I changes only very slightly with changes to the material constants, the predicted value of ΔK_{th} depends only on Young's modulus, Poisson's ratio and the surface energy of the material. Clearly, the model cannot distinguish between different compositions and heat treatments within a given alloy system, so it cannot describe the known threshold variations within each system. However, it was found that the equation did give an accurate prediction of the *lowest* threshold to be measured in any given alloy system. For example, the lowest result for any ferritic steel is about 2 MPa $\sqrt{m}$, and this is accurately reflected in the result from equation (2.1) if values for Fe are used. Figure 2.1 shows that the same applies for other classes of metals.

This observation lead to the concept that a simple model such as this might predict a lower bound for ΔK_{th}. This lower bound might apply in the absence of closure effects (i.e. at high R ratios) and in the absence of microstructural effects. For instance, the lowest thresholds in steels occur in very fine-scale microstructures; a very fine microstructure might be perceived as a homogeneous continuum, even under threshold conditions. Any increase in microstructural scale would then tend to increase the threshold away from this lower bound.

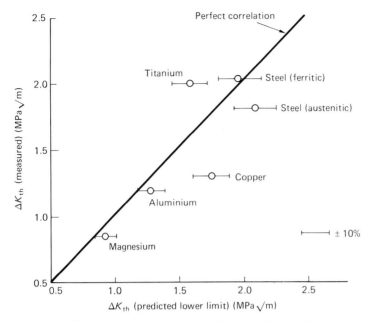

Figure 2.1 Comparison between predicted lower limits and lowest measured threshold values[2]

Extending the argument to include microstructural concepts, the author modified the plastic work term, making the assumption that plasticity always extended from the crack tip as far as the first grain boundary, even in cases where the predicted plastic zone size was much smaller than the grain size. This assumption is intuitively reasonable if grain boundaries are the principal barriers to slip, and is substantiated by reported observations of near-tip conditions. This leads to a model in which the plastic work term is: (a) larger than previously predicted; and (b) tending to increase with grain size. This modification leads to a surprisingly simple result from the energy balance:

$$\Delta K_{th} = \sigma_y \left[\frac{2.82\pi d}{(1 - v^2)} \right]^{1/2} \tag{2.3}$$

where d = grain size, measured as a two-dimensional linear intercept. This equation gave surprisingly good predictions for a whole range of materials (Figure 2.2), including bainitic and martensitic steels if d was replaced by lath width. In two cases a significant overprediction was traced to non-uniformity of grain size distribution.

The author does not propose that this energy-balance model is the solution to all our problems. In fact, looking back on it, the formulation contained some important errors in terms of the assumption of stress/strain conditions in the plastic zone. Undoubtably, the predictive success of the model is fortuitous, but its general principles seem sound, and serve to illustrate some general rules, for example:

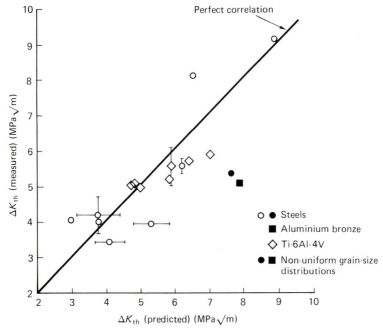

Figure 2.2 Comparison between predicted and measured threshold values for a variety of materials[2]

1. It seems likely that there exists a lower-bound threshold value, free from the effects of closure and microstructural inhomogenaeties.
2. One effect of microstructural features may be to control the physical extent of the plastic zone, thereby controlling the energy required for crack advance.

Criticisms of the Griffith approach

Guiu and Stevens[3] are among a growing band of theorists who believe that the Griffith approach, as modified by Irwin to introduce a plastic work term, is incorrect. Put simply, they suggest that the plastic term should be placed on the other side of the equation, as an energy-generating mechanism, not an energy-consuming one. This naturally leads to a considerable imbalance in the equation; at normal K_{Ic} values there would be far more energy available than required for crack advance, and this would still be the case even at threshold stress intensity values.

Guiu and Stevens argue that crack propagation is limited not by the availability of energy on a global scale, but by the need to attain certain conditions of stress or strain at the crack tip, to operate a crack advance mechanism. Crack propagation becomes a *kinetically* controlled event, rather than a thermodynamically controlled one. Guiu and Stevens advise that a more suitable condition for crack propagation is the need to exceed the cohesive strength of the material close to the crack tip; shear decohesion may normally be the easiest fracture mode.

Models based on dislocation dynamics

A number of workers, including Yokabori and Yokabori[4–9] and Mura and Weertman[10], have developed sophisticated models based on considerations of the behaviour of dislocations in the region ahead of the crack tip. The development of the Yokabori model is typical: initially, a model was derived, based solely on energy considerations for dislocation emission from the crack tip. Subsequently, this was modified by allowing the emitted dislocations to pile up at the first grain boundary and thus limit the outward flow of further dislocations from the crack tip. Thus, as for the energy models discussed above, a microstructural barrier is introduced, increasing the value of the predicted threshold. Further modifications considered a crack advance mechanism in which the role of the slip band was to create a microcrack at the grain boundary which eventually linked up with the main crack.

Similar arguments have been used by Mura and Weertman[10], Tanaka and Nakai[11], Sadananda and Shahinian[12] and Tanaka and Mura[13]. In the latter case, crack advance is envisaged to occur by decohesion of the slip band itself, at a given strain energy density. Chonghua and Minggao[14] also used a model based on critical strain range at the crack tip, but their model, though giving accurate predictions, depended critically on the value of the root radius for the blunted crack, which in practice cannot be accurately measured or predicted.

Generally speaking, the simpler dislocation-emission models tend to yield low values of ΔK_{th}, dependent on material constants such as E, γ and ν, in a fashion similar to the energy-based models discussed above. The result of Weertman and Mura is typical:

$$K = 0.6 \left[\frac{2\gamma E}{(1 - \nu^2)} \right]^{1/2} \tag{2.4}$$

In this case K is the limiting value of K_{max} in the cycle, so R ratio effects are incorporated to some extent.

The incorporation of the grain boundary barrier tends to increase the predicted thresholds to realistic values and to generate grain size dependences of the form:

$$\Delta K_{th} = A + B\sqrt{d} \tag{2.5}$$

where A and B are constants. Such a dependence is often found in practice for relatively simple microstructures such as low-carbon mild steels. Figure 2.3 shows typical results, from Taylor[15].

A criticism of such models is that they generally require input in the form of many basic material parameters which can often only be estimated very approximately and which critically affect the result. Also, it should be noted that there is no direct evidence for the mechanism of crack extension by linking of grain boundary microcracks, despite the number of electron-microscope observations of growing cracks, as described below.

It is an interesting historical development that dislocation-based theories of fatigue, which were once very popular before the advent of fracture-mechanics approaches to the problem, are now being reinvented

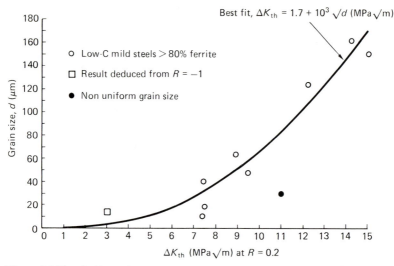

Figure 2.3 Threshold as a function of grain size in low-carbon mild steels[15]

to explain the events in the crack-tip process zone. Many of the above theories either directly invoke, or bear close resemblance to, concepts which were formally used to describe persistent slip band formation and stage I fatigue crack growth in plastic stress fields.

Put simply, many of the above models reduce to the concept (Figure 2.4) of a slip band reaching from the crack tip to a grain boundary, loaded by a stress equal to the cyclic yield strength of the material, σ_{yc}. Using the simplest fracture-mechanics equation and assuming the slip band to be a crack which will just be able to propagate at the threshold, gives the relationship:

$$\Delta K_{th} = \sigma_{yc} \sqrt{(\pi d)} \tag{2.6}$$

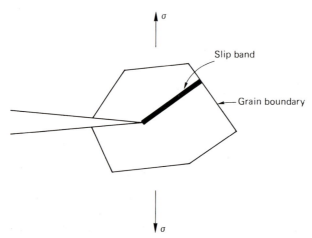

Figure 2.4 Slip band and grain boundary at the crack tip

This expression is clearly invalid for a number of reasons, not least for its use of an LEFM formula in a fully plastic stress field. However, the equation does in fact predict the observed dependence on grain size and yield strength for many materials. Similar, more complex, relationships are derived by many of the dislocation-based and plasticity-based theories mentioned above.

Two recent approaches[16,17] have yielded relatively simple theoretical developments which are substantiated by some experimental data. The blocked slip band model of Du Bai-ping, Li Nian and Zhou Hui-jiu[16] addresses quenched and tempered steels, postulating a situation in which the critical microstructural unit changes from martensite packet size (in undertempered steels) to prior-austenite grain size (in overtempered steels), with a consequent increase in ΔK_{th}. These effects are discussed further in Chapter 5.

In a similar vein, Mutah and Radhakrishnan[17] use an approach taken directly from the Hall–Petch model in order to predict threshold values for both stainless and low-carbon steels.

Models based on crack-tip plasticity

The distinction between this section and the preceding one is somewhat arbitrary; a number of models have been derived, which are concerned with the extent and magnitude of crack-tip plasticity without dealing directly with the behaviour of dislocations. The plastic zone is seen as a region in which fracture events take place, causing crack propagation.

Plastic zone size

A model which has had some predictive success relates plastic zone size in the near-threshold region to grain size. It is argued that, as ΔK is reduced from a high value, a change in fracture mechanism should be expected when the plastic zone ahead of the crack becomes confined in the region between the tip and the first grain boundary. It is reasonable to suppose that fracture events, such as shear yielding, will allow relatively easy crack propagation within the grain, with the critical event being the propagation of the plastic zone into the next grain ahead.

Yoder, Cooley and Crooker have advanced this approach in titanium alloys and steels in a number of well-defended publications[18–21]. They note also that secondary transitions, giving discontinuities in the growth-rate curve at higher ΔK values, may be associated with larger microstructural parameters.

Whatever the detailed explanation of the mechanism, this concept relates very well to the common experimental observation that the mode of crack growth changes at growth rates corresponding to the 'knee' of the growth-rate curve, i.e. the transition point between the Paris region and the near-threshold region. In many, but not all cases, this transition point also corresponds approximately to the value of ΔK at which the cyclic plastic zone size is equal to the grain size. This transition point, labelled ΔK_t, is not always clearly defined on growth-rate plots, but when it is there

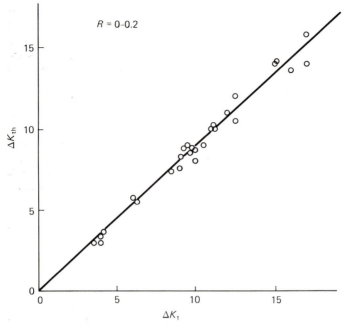

Figure 2.5 Comparison of ΔK_{th} and ΔK_t ('knee') values from crack growth-rate data on various materials

is a close correspondance between ΔK_t and ΔK_{th}, with the latter being typically 10 per cent lower (Figure 2.5). This relationship occurs because most materials demonstrate the same steep slopes in the near-threshold region.

Figure 2.6 shows that for many materials there is reasonable correlation between grain size and the function:

$$\left(\frac{\Delta K_{th}}{\sigma_y}\right)^2 \qquad (2.7)$$

which is proportional to the plastic zone size at $\Delta K = \Delta K_{th}$. In fact, this simple dependence is surprising, since it really ought to apply only to cases where the grain boundary is the principal barrier in the material. It would not be expected to be significant in, for example, high-strength aluminium alloys or nickel-base superalloys, where strength is controlled by precipitation. Further, in materials with coarse, two-phase microstructures such as ferrite/pearlite and ferrite/martensite, one might expect the significant distance to be the separation of the colonies of the hard phase. In fact, the theory works well for mild steels, even with quite high pearlite contents.

This approach enjoyed considerable popularity some years ago, before the advent of crack closure theory, which has since tended to displace it. It is recommended that the approach be re-examined in an attempt to place it on a surer footing. Two elements of the theory merit further discussion:

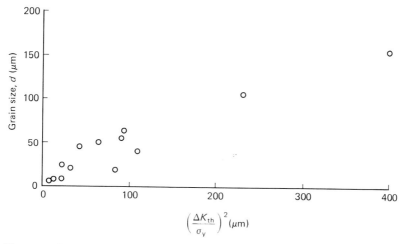

Figure 2.6 Comparison of grain size, d, and the function $(\Delta K_{th}/\sigma_y)^2$ for various materials

1. The microstructural aspects would benefit from more careful definition, including the recognition that a grain is a three-dimensional entity and that many grains of different sizes exist along the crack front. More will be said of this problem below.
2. The model should incorporate a definite argument as to the means of crack extension in the plastic zone; information may be brought in from many sources, including classic low-cycle fatigue mechanisms on the one hand and recent microscopic observations of crack growth on the other (see below).

Low-cycle fatigue at the crack tip

Models which are in a sense derived from the above approach have been developed by Starke et al. [22] and Chesnutt [23].

Basically these models treat the region ahead of the crack as a low-cycle fatigue specimen. Crack growth rates and thresholds are dependent on $\Delta\epsilon_{pl}$, the local range of cyclic plastic strain. Both models are derived from an earlier attempt by Chakrabortty [24]: a microstructural parameter is included as an influence on the local plastic strain field. The models tend to predict too high a value for the threshold when compared with experimental results, possibly due to the difficulty of estimating the strain distribution close to the crack-tip with any accuracy.

Crack-tip opening displacement

Another similar approach is to assume that crack propagation is related to crack-tip opening displacement [25], CTOD, where:

$$CTOD = \frac{K^2}{\sigma_y E} \tag{2.8}$$

This is similar to the low-cycle fatigue approach in that it assumes that material immediately ahead of the crack tip is being subjected to a specific cyclic strain. The threshold may occur if there is just enough strain available to fracture the crack-tip element of material. Rearrangement of the above equation, with CTOD fixed as a material constant for the threshold condition, leads to a prediction in which ΔK_{th} is proportional to $\sqrt{(E\sigma_y)}$, which efficiently predicts the general variation of threshold from one class of material to another[24].

The problem of the crack front

A general criticism of all the above mechanistic models is that they fail to recognize that the crack tip is not a single point but a line, the crack front. Any condition for crack advance which involves the interaction of the crack, or its plastic zone, with the microstructure ahead will naturally encounter the following problem: different parts of the crack front will find it easier or harder to advance, owing to the varying condition of the local microstructure along the crack front. The same problem is encountered in other fields, for instance the motion of a dislocation through a matrix containing randomly placed precipitates.

Some condition has to be applied, presumably allowing parts of the crack front to advance while other parts do not, giving an irregular, locally bowed shape to the crack front which will alter the local stress, strain and energy conditions. Observations of growing cracks often reveal periods of dormancy between spurts of growth; this may be simply because one is looking at one point (inevitably a surface) of a crack front, some of which is moving and changing the conditions at the point of observation until they become favourable for growth.

This concept has been considered by Gerberich and Moody[26] and by Edwards and Martin[27], who define a 'semicohesive zone' (Figure 2.7) containing ligaments of unfractured material. In the case of Edwards and Martin, the effect of dispersoid additions to an Al alloy, which increased ΔK_{th}, was explained by strengthening of these ligaments, which failed by a

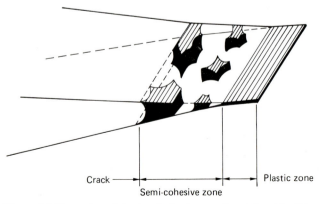

Crack
Semi-cohesive zone
Plastic zone

Figure 2.7 Illustration of the 'semicohesive zone' postulated by Edwards and Martin[27]

low-cycle fatigue process. As we have shown above, the effect of the dispersoids could equally be explained in terms of an increase in yield strength at constant grain size.

The effect of grain size, or microstructural unit size, might also be incorporated into this ligament model by assuming that the ligament size (i.e. cross-sectional area) is related to the grain size.

This would seem to be an appropriate application for a computer simulation, incorporating a crack front and a series of grain boundaries and other microstructural features. Such a simulation would enable one to investigate various conditions for crack-front movement.

Crack branching and tortuosity

A problem which has been addressed by Suresh and co-workers[28–30] and Gray, Thompson and Williams[31] is the fact that a real crack is generally not a straight line but a tortuous shape which may involve sudden deflections through large angles as well as branches (see Figure 2.8). This has consequences not only for crack closure (see Chapter 3) but also for the value of K at the tip. Suresh showed that relatively small deflections from linearity at the crack tip caused significant reductions in effective ΔK, and he has made some attempts at the difficult problem of modelling complete cracks with varying degrees of deflection.

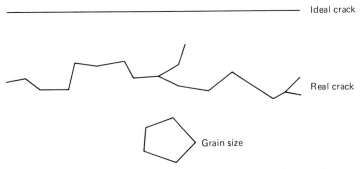

Figure 2.8 Ideal crack and real crack, showing branching and tortuosity

In practice, some material conditions give rise to very great angular deflections (Figure 2.9) while others do not, and this may be reflected in significantly higher thresholds for the deflected cracks. This is a particularly active method of improving fatigue behaviour in aluminium alloys and nickel-base alloys (see Chapter 5), and in other precipitation-hardened systems; changes to precipitate condition can be used to alter the degree of crack deflection dramatically.

The effect of crack deflection on stress intensity and on closure level is very difficult to quantify since it involves a statistical assessment of random angular deflections and random asperity contacts. Inevitably, these effects interact with other mechanisms such as those listed above, tending always to increase the measured threshold value above what might be predicted for a simplified straight crack.

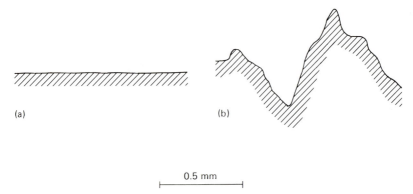

0.5 mm

Figure 2.9 Cross-sections of crack growth paths near threshold, taken from Petit[32]; (a) 7075 highly overaged, in air or vacuum; (b) 7075-T351 in vacuum

Direct observations of growing cracks

A number of workers, notably Lankford and Davidson[33,34] and Morris and James and Morris[35,36] have successfully observed the behaviour of cracks when cyclically loaded while mounted in a scanning-electron-microscope. Such observations are naturally of great interest to those attempting to develop mechanistic theories of crack growth.

Parameters which have been measured include crack-tip opening displacement, local strain and crack-closure stress. Also the interaction between a growing crack and a grain boundary can be observed. Results to date are still difficult to correlate, and different workers tend to reach different conclusions, but the observations yield valuable insights of a qualitative and semiquantitative nature. An inevitable problem is that the specimens involved tend to be small, and that the crack tip can only be viewed at the surface, where plane-stress conditions prevail.

Other workers have used brittle polymeric materials, which have two advantages: the crack front within the specimen can be viewed directly if the material is transparent, and stress information can be obtained using photoelastic techniques. Crack propagation and closure in fatigue can be reproduced in this way, but results are of dubious benefit since the mechanisms of crack growth and plastic deformation are not the same in polymers as they are in metals. Therefore, any information on stress fields or closure conditions generated using these materials should be treated with caution.

Modelling the shape of the growth-rate curve

It has been noted by a number of workers that the 'knee' or transition point at which the growth-rate curve deviates from linearity towards the threshold always occurs at a growth rate da/dN approximately equal to the atomic lattice spacing of the material[37]. This is reasonable, since to attain lower growth rates the crack front must necessarily advance discontinuously, and this may of itself demand a change in growth mechanism.

It is tempting, therefore, to derive threshold values using only the Paris constants for the straight-line region and the da/dN value for the transition, assuming some form for the asymptotic near-threshold part of the curve. Unfortunately the transition point is often difficult to define and rarely occurs exactly as predicted, and the asymptotic behaviour is sufficiently variable to introduce large errors in the final prediction of ΔK_{th}, however, it is possible that this concept could be built on successfully.

One approach is to combine the lattice spacing concept with a mechanistic theory of crack advance, as done by Mura and Weertman, for example [10].

Other workers have attempted to analyse the data of the growth-rate curve in the hope of finding trends which allow prediction in the absence of mechanistic theories. The work of Radhakrishnan [38] is typical. Noting that in the Paris region, data corresponds to:

$$\frac{da}{dN} = B\Delta K^n \tag{2.9}$$

and assuming that the near-threshold region corresponds to:

$$\frac{da}{dN} = C\Delta K^m \, (m \gg n) \tag{2.10}$$

Radhakrishnan attempts to show that the four constants B, C, n and m are related by equations of the form:

$$\left.\begin{array}{l} \log(B) = \log(B_{\text{o}}) - nq \\ \log(C) = \log(C_{\text{o}}) - mp \end{array}\right\} \tag{2.11}$$

where B_{o}, C_{o}, p and q are constants for a given alloy system. The consequence of this is that all the lines formed by equation (2.7), if extrapolated, should cross at some unique point (which is at a high value of ΔK); likewise all the lines from equation (2.8) should meet at some other point. These 'pivot points' are not given any mechanistic significance in Radhakrishnan's treatment. Bailon, Masounave and Bathias [39] have also analysed growth-rate data in a similar fashion.

Such approaches may be valuable as data-handling exercises which, if they are able to detect trends in the variation of the growth-rate curve, may be used as design rules. They are particularly useful in revealing any unusual materials which, for some reason or other, do not conform to a general pattern. However, one remains sceptical of any effects for which no mechanistic basis can be demonstrated.

The effect of Young's modulus

A number of studies have noted the strong effect of Young's modulus on fatigue crack growth behaviour. As a general rule, ΔK_{th} can be said to be proportional to E, in the sense that data from many different alloy systems tend to gather in a single scatter band if da/dN is plotted against $\Delta K/E$

24

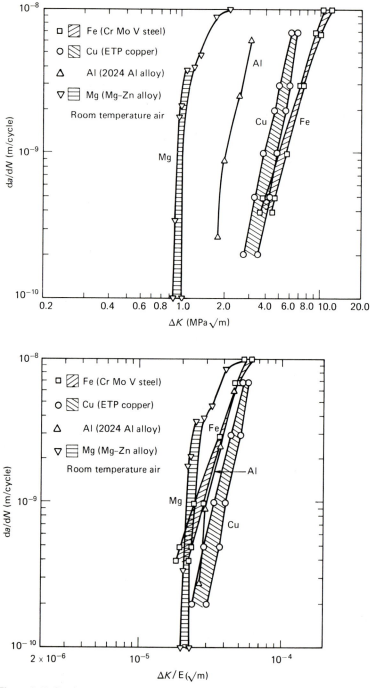

Figure 2.10 Crack growth-rate plot for various materials, using ΔK/E; Liaw[40]

(Figure 2.10). This has been shown, for example, by Liaw, Leax and Logsden[40].

Concluding remarks

This chapter should be read in conjunction with the following one on crack closure, since closure now constitutes a major mechanism and has generated a number of attendant theories, and also because most of the above models must necessarily be modified to take closure concepts into account.

I shall conclude this chapter by noting the comment of a speaker at a recent fatigue conference, who referred to a 'shopping mall' of mechanisms. The current feeling is that so many mechanistic theories have been advanced to explain near-threshold behaviour that any experimental data can be explained away by reference to one or another of them. A considerable work of assimilation and comparison is required if one is to decide which approach is valid under which set of conditions.

References

1. Purushothaman, S. and Tien, J. K. (1978) *Materials Science and Engineering,* **34**, 241
2. Taylor, D. (1982) In *Fatigue Thresholds,* EMAS, Warley, UK, p. 5
3. Guiu, F. and Stevens, R. N. (1986) In *The Behaviour of Short Fatigue Cracks* (EGF1), MEP, Bury St Edmunds, UK, p. 407
4. Yokabori, A. T. (1978) *International Journal of Fracture,* **14**, R317
5. Yokabori, A. T. (1978) *International Journal of Fracture,* **14**, R315
6. Yokabori, A. T. and Yokabori, T. (1981) In *Advances in Fracture Research* (ICF5) Pergamon, Oxford, p. 1373
7. Yokabori, T., Yokabori, A. T. and Kamei, A. (1974) *Philosophical Magazine,* **30**, 367
8. Yokabori, T., Yokabori, A. T. and Kamei, A. (1975) *Journal of Applied Physics,* **46**, 3720
9. Yokabori, A. T. and Yokabori, T. (1982) In *Fatigue Thresholds,* EMAS, Warley, UK, p. 171
10. Mura, T. and Weertman, J. (1984) In *Fatigue Crack Growth Threshold Concepts,* TMS–AIME, USA p. 513
11. Tanaka, K. and Nakai, Y. (1984) In *Fatigue Crack Growth Threshold Concepts,* EMAS, Warley, UK, p. 497
12. Sadananda, K. and Shahinian, P. (1977) *International Journal of Fracture,* **13**, 585
13. Tanaka, K. and Mura, T. Work reported in reference 11
14. Chonghua, Y. and Minggao, Y. (1979) *Fatigue of Engineering Materials and Structures,* 189
15. Taylor, D. (1985) In *Fatigue 84,* EMAS, Warley, UK, p. 327
16. Du Bai Ping, Li Nian and Zhou Hui-Jiu (1987) *International Journal of Fracture,* **9**, 43
17. Mutah, Y. and Radhakrishnan, V. M. (1981) *Transactions of the American Society of Mechanical Engineers, Journal of Engineering Materials and Technology,* **103**, 229
18. Yoder, G. R., Cooley, L. A. and Crooker, T. W. (1980) In *Titanium '80* (eds. Kimura and Izumi) p. 1865
19. Yoder, G. R., Cooley, L. A. and Crooker, T. W. (1982) *Scripta Metallurgica,* **16**, 1021
20. Yoder, G. R., Cooley, L. A. and Crooker, T. W. (1983) In *Fracture Mechanics, Fourteenth Symposium – Vol. I: Theory and Analysis,* ASTM STP791, pp. 1–348

21. Yoder, G. R., Cooley, L. A. and Crooker, T. W. (1985) In *Fatigue 84,* EMAS, Warley, UK, p. 351
22. Starke, E. A., Lin, F. S., Chen, R. T. and Heikkenen, H. C. (1984) In *Fatigue Crack Growth Threshold Concepts,* TMS-AIME, USA, p. 43
23. Chesnutt, J. C. and Wert, J. A. (1984) In *Fatigue Crack Growth Threshold Concepts,* TMS-AIME, USA, p. 83
24. Chakrabortty, S. B. (1979) AIME Annual Meeting, New Orleans
25. Beevers, C. J. (1981) In *Fatigue Thresholds,* EMAS, Warley, UK, p. 17
26. Gerberich, W. W. and Moody, N. R. (1979) In *Fatigue Mechanisms,* ASTM STP675, p. 295
27. Edwards, L. and Martin, J. W. (1983) *Metal Science,* **17,** 511
28. Suresh, S. (1983) *Engineering Fracture Mechanics,* **18,** 577
29. Suresh, S. (1984) Report MRL E-153, Brown University, USA
30. Suresh, S. (1983) *Metallurgical Transactions,* **14A,** 2375
31. Gray, G. T., Thomspon, A. W. and Williams, J. C. (1984) In *Fatigue Crack Growth Threshold Concepts,* TMS-AIME, USA, p. 131
32. Petit, J. (1984) In *Fatigue Crack Growth Threshold Concepts,* TMS-AIME, USA, p. 3.
33. Lankford, J. (1984) In *Fatigue Crack Growth Threshold Concepts,* TMS-AIME, USA, p. 447
34. Davidson, D. L. and Lankford, J. (1980) *Fatigue of Engineering Materials and Structures,* **3,** 289
35. Morris, W. L. and James, M. R. (1984) In *Fatigue Crack Growth Threshold Concepts,* TMS-AIME, USA, p. 479
36. Morris, W. L. *Metallurgical Transactions,* **11A,** 1117
37. Pook, L. P. and Frost, N. E. (1973) *International Journal of Fracture,* **9,** 53
38. Radhakrishnan, V. M. *Engineering Fracture Mechanics,* **13,** 129
39. Bailon, J-P., Masounave, J. and Bathias, C. (1977) *Scripta Metallurgica,* **11,** 1101
40. Liaw, P. K., Leax, T. R. and Logsden, W. A. (1983) *Acta Metallurgica,* **31,** 1581

3 Crack closure

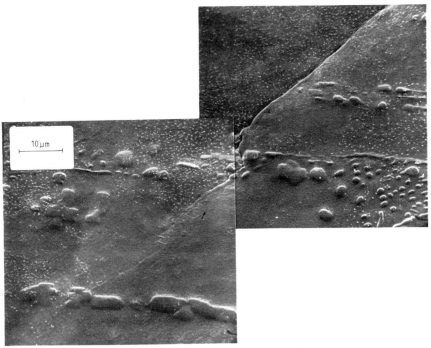

10 μm

Closure near the tip of a fatigue crack – scanning electron replica micrograph

Introduction

The phenomenon of crack closure is now widely acknowledged to exert a strong influence on near-threshold fatigue crack growth, and the volume of literature generated on closure is large enough to merit a separate chapter in this book. The various other mechanisms and models used in near-threshold studies are covered in Chapter 2.

The concept of closure has caused nothing less than a revolution in our thinking about the behaviour of fatigue cracks. Closure ideas have helped to explain, at least in a qualitative manner, many crack propagation effects, especially concerning the near-threshold region, such that some understanding of closure concepts is now essential for anyone handling fatigue problems.

Closure: basic outline

Consider a specimen containing a crack, experiencing a fatigue cycle with a mean load of zero. During the compressive half of the cycle, it is reasonable to assume that the crack is closed, and transmits compressive stress across its faces. Therefore, in some sense the specimen can be described as a cracked body during the tensile part of the cycle, and an uncracked body during the compressive part. One may go further, and assume that the tensile part of the cycle is the only *effective* part, and therefore assume that the body is experiencing an effective stress range equal to only half the applied stress range. This creates an effective stress intensity range, denoted ΔK_{eff}. This concept of an effective range of ΔK, defining that part of the cycle for which the crack is open, is the basis of crack closure.

It must be stressed that this assumption of an effective stress range is a relatively inaccurate, simplifying assumption. We are well aware that the compressive part of the loading cycle *does* have effects at the crack tip, such as crack-tip sharpening and changes to the plastic zone, which aid crack propagation in the next tensile half-cycle. If the compressive part was completely redundant one would find that behaviour was controlled by maximum stress only for all negative R ratios, whereas in fact this is not the case (see Chapter 7). We will return to this problem at a later stage.

Now, it is found in practice that it is common for a crack to be closed, even during the tensile part of the cycle, and even if the loading is entirely tensile (i.e. with an R ratio greater than zero). The situation is illustrated in Figure 3.1, which gives schematic plots of stress intensity and crack opening displacement for tension/tension cycling for a typical metal. Two points can be defined on the K cycle, at which the crack begins to close (on the decreasing leg of the cycle) and open (on the increasing leg) respectively. These two points may occur at the same K value, but commonly there is a lag effect, the value of K_{op} being somewhat higher than K_{clo}. This gives some latitude to the choice of the value of ΔK_{eff}, but it is generally taken as:

$$\Delta K_{eff} = K_{max} - K_{clo} \tag{3.1}$$

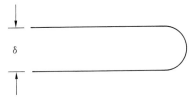

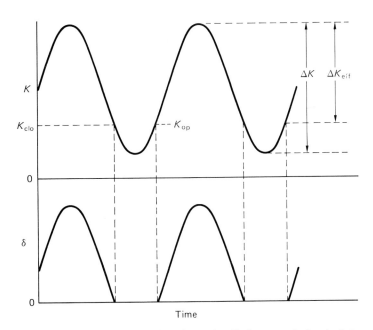

Figure 3.1 Schematic variation of crack opening displacement during the fatigue cycle; definition of crack closure terms

using the point at which the crack is deemed to be 'fully closed' and therefore capable of transmitting compressive stress in the same manner as intact metal. One further parameter can be defined from this measurement: the closure ratio, U:

$$U = \frac{\Delta K_{\text{eff}}}{\Delta K} \tag{3.2}$$

This gives the portion of the cycle over which the crack is opening; it therefore tends to unity for cracks which do not show closure, and tends to zero if closure is complete throughout the cycle. It should be noted, however, that some authors define U differently, giving an inverse relationship to the one defined here.

Mechanisms of closure

Why does closure occur during the tensile part of the cycle? There are a number of reasons. The simplest cause is a blockage caused by ingress of debris into the crack, or of a viscous fluid, or a blockage caused by corrosion products which form on the crack faces. An important example of the latter is the build-up of a voluminous oxide layer, accentuated by fretting action between the opposing crack faces of, for example, an alloy steel.

Also if the crack faces are at all rough, i.e. if the fracture path is not completely planar, then any relative movement during the opening part of the cycle may result in a poor fit between the two surfaces during the closing part. Asperities will come into contact and begin to transmit stress, effectively causing closure.

Finally there is a more fundamental source of closure which occurs in almost all materials at low R ratios. This arises from the nature of the plastic zone itself. In addition to the plastic zone in front of the crack, there is necessarily a 'wake' of material along the crack sides, the remainder of the plastic zone as the crack travels forwards (Figure 3.2). This plastic

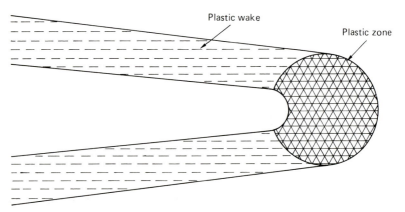

Figure 3.2 Illustration of the plastic wake, which causes closure through the action of residual stresses

wake is a source of compressive residual stress. To appreciate this, consider the analogy of a metal surface which has been hardened by shot peening or hammer peening. In this case the plastic strain induced by the peening process is unable to be realized as a surface expansion, owing to the constraints of the material below the surface. It therefore manifests itself as a compressive residual stress. In the same way, material in the plastic zone of a crack is constrained by the elastic material around it, so residual stress results, the effect of which is to tend to press the sides of the crack together. This effectively imposes a constant stress on top of the applied cyclic stress, shifting the mean stress of the cycle to a lower value. Therefore, if the applied cycle is at a sufficiently low R ratio the residual stress will bring part of the cycle into the compressive region, creating closure.

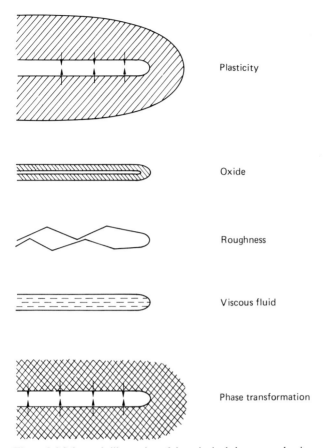

Plasticity

Oxide

Roughness

Viscous fluid

Phase transformation

Figure 3.3 Schematic illustration of the principal closure mechanisms

This fundamental source of closure was first conjectured and later established experimentally by Elber[1,2]. Some years passed before it was appreciated that other mechanisms were also causing closure. To date, five different closure mechanisms have been established; these are illustrated schematically in Figure 3.3 and discussed in detail as follows:

Plasticity-induced closure

This name is now given to the basic Elber mechanism described above. It is found to be very important in some materials at low and negative R ratios. The value of K_{clo} is generally not constant with R, unfortunately, and its value is very alloy-specific, possibly owing to changes in cyclic stress/strain characteristics. This type of closure acts to a greater degree in plane-stress conditions than plane strain, so that it is more prevalent in thin specimens and at the edges of thicker specimens. This has been elegantly demonstrated by Blom and Holm using a finite-element computer model of the crack tip[3] by which they were able to predict closure values in an aluminium alloy[4].

Oxide-induced closure

This was originally observed in steels by Ritchie, Suresh and co-workers[5–7] who noticed that oxide deposits on fracture surfaces tended to be more pronounced near threshold and at low R values, causing black bands on a fracture surface which delineate the position of the crack front under near-threshold loading. Figure 3.4 illustrates this phenomenon. Measurements of the thickness of such oxide layers showed them to be of the same order of magnitude as the calculated crack-tip opening displacement near the threshold; Figure 3.5 shows how the measured thickness of the oxide layer increases at low R ratios, and how this coincides with an increase in the measured threshold value.

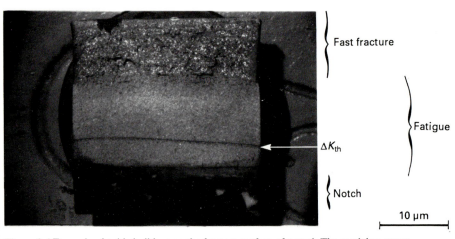

Fast fracture

Fatigue

ΔK_{th}

Notch

10 μm

Figure 3.4 Example of oxide build-up on the fracture surface of a steel. The crack has grown from bottom to top in the picture; arrows indicate the position of the crack at ΔK_{th}

It is important to note that the oxide layer can only form if closure is already occurring as a result of another mechanism, since fretting contact is required. It is thus only a means of enhancing existing closure levels, though in this role it may come to dominate the situation. It should also be noted, as has been discussed elsewhere, that, being basically a tribological effect, oxide closure is affected strongly by temperature and loading frequency (see Chapters 4 and 6).

The concept can be extended to include any corrosion product produced at the crack faces, or any foriegn debris which may enter the crack from the external environment, although few practical studies exist to demonstrate the effectiveness of these other sources.

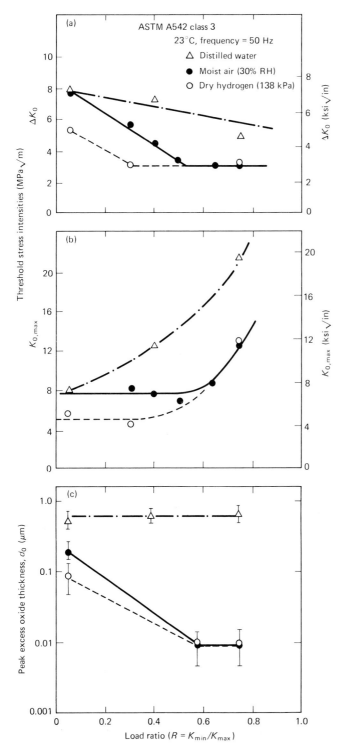

Figure 3.5 Oxide thickness measurements by Ritchie and Suresh[18]

Roughness-induced closure

Originally termed 'non-closure' by Beevers, this was historically the second closure mechanism to be demonstrated. While the crack is open, relative motion of the two faces can occur by the operation of mode II and III-type stresses; in particular it has been established that significant mode II stresses operate at the crack tip even under far-field mode I loading. If any such relative motion occurs, then the two halves of the crack will not fit together perfectly on closure. Opposing asperities will come into contact at some point and, since these can transmit stress across the crack faces, some degree of closure will occur.

It is difficult in this case to specify what quantitative effect this will have, in terms of the K_{clo} and ΔK_{eff} parameters, since the closure is necessarily occurring at isolated, random points. Suresh[8] has attempted to model this difficult situation, with some predictive success. There is a small error in the mathematics of this model, in that Suresh assumes the range of COD to be equal to K^2, when it fact it should be written as$(K_{max}^2 - K_{min}^2)$. This in fact leads to an underestimate of the amount of closure involved. Also, Suresh takes an idealized rough surface similar to that drawn in Figure 3.3, i.e. one consisting of regular planes at relatively small angles to the horizontal. Other workers have also modelled the effects of roughness and crack deflection[9–11].

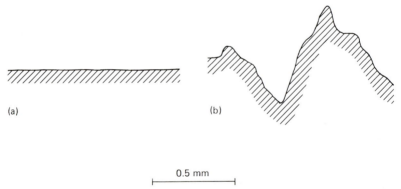

(a)   (b)

0.5 mm

Figure 3.6 Cross-sections of crack paths in two different heat treatments of an aluminium alloy, from Petit[12]

In fact, real cracks show surprisingly high angles of roughness; Figure 3.6 shows drawings of crack surfaces taken from results by Petit[12], comparing two heat treatments of the same aluminium alloy. The contrast between the smooth and rough surfaces is striking, and the rough surface shown in this case is not exceptional. Clearly, it only requires a small amount of relative movement to cause high levels of roughness-induced closure. Figure 3.7 shows asperity contact in a micrograph of a fatigue crack in a titanium alloy[13].

Generally speaking, the scale of the roughness is found to be related to the material grain size, because the path of the near-threshold fatigue crack tends to be relatively planar within each grain, with deflections at the grain

Figure 3.7 Asperity contact between crack faces in a titanium alloy, from Beevers[13]

boundary. This implies a possible grain size effect, since larger grain sizes will cause rougher surfaces. This effect is no doubt present, but consideration of the geometry of the situation will show that the *angle* of roughness is much more important than the scale. Indeed, in theory, if the grain size is much larger than the COD (which it generally is) then the opening and closing of the crack will take place on a much smaller scale than the roughness (Figure 3.8) and grain size would therefore be unimportant.

Some very interesting results have been obtained by McEvily and Minakawa[14] who tested a material with an ultra-fine grain size of 0.2 μm. They measured no closure at all, and recorded no R ratio effect. It is not

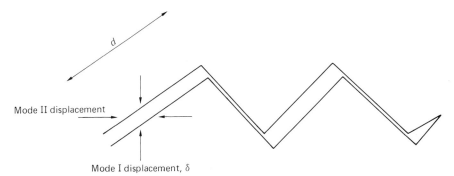

Mode II displacement

Mode I displacement, δ

Figure 3.8 Schematic illustration of crack opening and roughness

clear why plasticity-induced closure played no part here, but the results certainly provide food for thought.

A common source of roughness is change of slip character; for example underaged aluminium alloys show very rough surfaces, and correspondingly good near-threshold fatigue behaviour (see Chapter 5). Other sources of roughness include slip steps[15], which would be effective even without any mode II motion of crack faces. Environmental interaction also changes roughness, especially if it induces, or supresses, intergranular growth.

An unusual example of excessive crack deflection occurs in the work of Gregory[16] on magnesium alloys. It was shown that an Mg–Li alloy showed strikingly different crack paths from those seen in pure Mg, due to the availability of prismatic slip systems in the alloy. Another unusual source of roughness-induced closure is found in the work of Dauskardt *et al.*, testing A533B pressure-vessel steels. In this case the introduction of microvoids on grain boundaries caused a large reduction in K_{Ic}, but only a very small reduction in ΔK_{th}. Roughness-induced closure had been greatly increased by the presence of voids on the fracture surfaces.

Some workers have employed standard surface roughness measurements, using a stylus instrument, to characterize the susceptibility of a fracture surface to roughness-induced closure[12,17]. Again it should be noted that such a measurement does not generally include the angle of roughness, or any measure of the horizontal scale (as opposed to the vertical height) of the roughness. The extensive work in this area from the field of tribology could usefully be accessed in developing this approach further.

In summary it is clear that roughness-induced closure is important, and arises from a wide range of causes. It is also clear from the geometry of the situation that it is difficult to predict the onset of this form of closure, and even more difficult to predict its effect on crack growth behaviour.

On the other hand, this mechanism offers great opportunities for improved alloy design with the aim of encouraging roughness, either by altering slip character or by arranging phases in coarse, multiphase microstructures (see Chapter 5).

Viscous-fluid-induced closure

This mechanism, demonstrated by Suresh, Ritchie and co-workers[18] relies on the fact that a viscous fluid is capable of transmitting stress. Therefore a crack filled with, say, oil, may demonstrate some closure effects at all stress levels, dependent on the loading frequency. This is an interesting mechanism, but confined to fluids which are viscous enough to exert an effect but also capable of flowing into the narrow openings afforded by near-threshold cracks.

Phase-transformation-induced closure

It was also noted by Suresh and Ritchie[18] that closure might arise from a material which shows a stress-induced phase transformation, the phase formed being more voluminous than the original phase. In principle, the mechanism is available to a number of materials, including stainless steels and certain polymers and ceramics[19], but no material has yet been designed specifically to take advantage of this mechanism.

The measurement of closure

Mechanistically, as we have shown above, the closure story is a very complex one. In practice, one can measure the overall degree of closure, deriving values for K_{clo} and ΔK_{eff}, as defined above, whatever mechanism or combination of mechanisms happen to be operating. I propose, therefore, to discuss the measurement of closure, and of closure-related phenomena, in some detail. I propose also to suggest a reason why the standard method of measurement of closure, using compliance, has proved so useful in describing crack propagation phenomena.

Compliance measurement

The simplest, and most generally used method of measuring closure is to measure specimen compliance. Since compliance depends on crack length, then a load/deflection plot for the specimen should show a change in slope, corresponding to the onset of closure. Figure 3.9 shows ideal behaviour, assuming that closure sets in at one specific point, that $K_{clo} = K_{op}$, and that the compliance below the closure load is the same as the compliance of the uncracked specimen. In principle, any deflection parameter of the

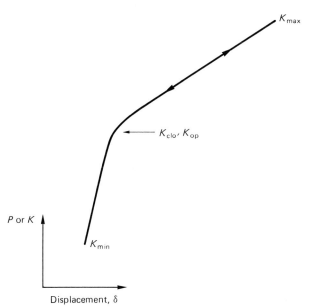

Figure 3.9 Idealized load/deflection trace, illustrating closure

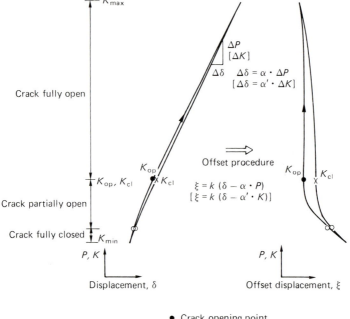

Figure 3.10 Actual load/deflection trace, showing the enhancement of closure detection using a derivative analysis, from Fleck[17]

specimen can be used to define compliance, but results are better taken either from a crack-mouth clip gauge or a back-face strain gauge. Some workers have also used strain gauges mounted on the side of the specimen close to the crack tip.

Figure 3.10 shows an actual compliance curve, from Fleck[17]. In general the change in slope is relatively small and occurs gradually. For this reason a method was developed for taking the derivative of the line electronically, giving an output of slope versus load (also shown in Figure 3.10), which enhances the change in slope, allowing better definition of K_{clo}. Other problems include hysteresis (i.e. K_{clo} not equal to K_{op}) and signal noise; in many cases the load/displacement signals are much less 'clean' than Fleck's, giving rise to doubts as to the accuracy of the definition of K_{clo}. At present there is a considerable need for a comparative study of this method of measurement and the development of standards to cover its use.

Fleck[17] also developed a method for measuring closure in the centre of a specimen, i.e. under plain-strain conditions, using an elegantly designed instrument involving two piston rods which located on the upper and lower faces of the crack close to its tip. He was able in this way to demonstrate that plane-strain closure was finite, but considerably smaller than plane-stress closure.

Relationship between compliance variation and fatigue behaviour

Despite the relative inaccuracy of this method, values of K_{clo} and ΔK_{eff} gathered from it have been very successfully used by many workers. As will be shown below and in other chapters of this book, the use of ΔK_{eff} instead of ΔK is very valuable in understanding the R ratio effect and the behaviour of short cracks and notches, as well as rationalizing microstructural changes which affect crack growth rates. This is somewhat surprising considering the simplifying assumptions necessary in the adoption of the ΔK_{eff} parameter, and the inaccuracies of practical measurement. I propose to show that the compliance method of measurement is in fact measuring a more fundamental parameter, namely the elastic energy available for crack propagation.

Consider the form of the load/deflection plot, assuming ideal closure conditions, i.e. closure which occurs piecemeal at one value of load, such that the specimen containing the closed crack has the same compliance as an uncracked specimen. Behaviour before and after closure is taken to be linear elastic. Figure 3.11 shows the idealized plots for crack lengths 'a' and '$a + da$', da being an infinitessimal increase in crack length. For comparison, the lines for a similar specimen showing no closure are also drawn. Note that the two lines for length 'a' are parallel, likewise the lines for '$a + da$'.

It is normal to calculate the elastic energy lost during crack extension for the 'fixed grips' case, in which the specimen is loaded up to a load P, and then the crack allowed to extend by 'da'. The energy available for growth is then proportional to the area between the 'a' and '$a + da$' lines.

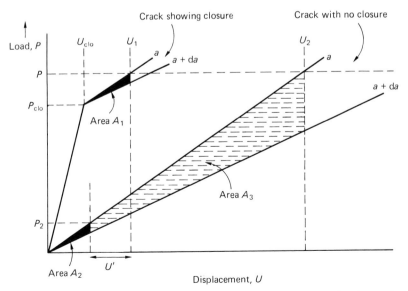

Figure 3.11 Idealised load/deflection plots for a crack subject to closure, showing crack lengths a and $(a + da)$.

Note now that the shaded areas A_1 and A_2 on the diagram are equal, U' being the value of U at which closure occurs. For identical loading conditions, the energy available is given by A_1 in the case of the 'closing' specimen, and $A_2 + A_3$ in the case of the specimen which does not close. In effect, we can say that, by ignoring closure, we are overestimating the available energy by A_3, this being the difference between the two areas.

Looking now at the load axis, and noting that, for constant crack length and geometry, P is proportional to K, we are saying that the closing specimen, loaded between zero and P, is in fact behaving as if it were a non-closing specimen loaded only from zero to P_2. By simple geometry, P_2 is equal to $P - P_{clo}$, which defines ΔP_{eff}, proportional to ΔK_{eff}.

To conclude, this definition of ΔK_{eff} is effectively a measure of the (reduced) energy available for crack propagation. As such, it would be an appropriate parameter to use, *whether or not* it corresponds to the physical process of crack closure. In real situations the plots corresponding to Figure 3.11 will be curved, largely because closure occurs gradually at different points along the crack length, but the definition of ΔK_{eff} will still be an approximate measure of the area under the relevant portion of the curve.

It would be an interesting exercise to re-evaluate compliance curves with this approach in mind. It should be pointed out that the above analysis in no way denies the importance of closure; it merely attempts to explain the relative success of the obviously simplistic parameter that is used to define the phenomenon.

Other methods of closure measurement

Potential drop

Electrical potential drop techniques, already well established for crack length measurement (see Chapter 4) could in principle be useful for detecting closure. In practice, results tend to show high scatter, hysteresis, and poor reproducibility (Figure 3.12). This may be because a small amount of asperity contact, occurring some way behind the crack tip, has a strong effect on the potential drop reading, while being relatively unimportant as a contribution to closure. Further work is recommended, however, to attempt to develop this system.

Acoustic measurements

Buck, Rehbein and Thompson[20] have used an acoustic method to detect and quantify asperity contact. They have also incorporated the method into a mechanical model which attempts to define an effective stress intensity in the presence of asperity contact using a somewhat more advanced approach than the simple ΔK_{eff} as defined above.

Direct Observation

Useful insight into closure can be obtained by observing acetate replicas of cracks, or even by viewing them directly through a microscope. Accuracy

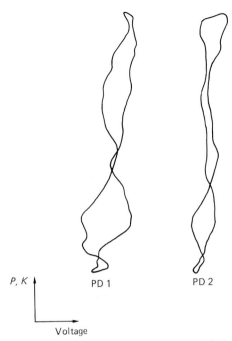

P, K

PD 1   PD 2

Voltage

Figure 3.12 Potential drop traces, from Fleck[17]

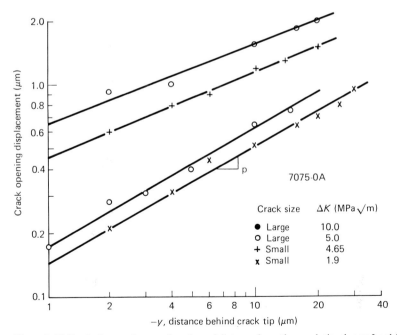

Figure 3.13 Crack-tip opening as a function of distance from the crack tip, due to Lankford and Davidson[21]. Note the logarithmic scale

of measurement is small, however, and it must be remembered that it is of necessity a surface which is being observed, and therefore a plane-stress region.

The use of *in-situ* straining stages inside electron-microscopes has enabled high-magnification observation of the same phenomenon[21,22]. In particular it is evident from these results, an example of which is shown in Figure 3.13, that crack opening remains small even at relatively large distances from the crack tip. This emphasises the importance of even small asperities or oxide debris over quite a large area of the local crack face.

The effect of closure on threshold values

Threshold values are influenced by many variables, including microstructure, environment and load characteristics. These effects will be discussed in detail in later chapters, but it is appropriate at this stage to summarise the role which crack closure plays in these areas.

R ratio

Threshold values in most materials tend to decrease markedly with increased R ratio in both the negative and positive ranges of R. Closure certainly exerts a strong influence, often a controlling influence, as evidenced by the fact that ΔK_{eff} values at the threshold are often constant for varying R ratios, at least up until the very high R values.

Microstructure

Changes to microstructure and mechanical properties of materials will alter threshold values for a variety of reasons. The principal way to increase closure in this case is to increase roughness-induced closure by encouraging crack deflection. This can be done by changing slip character or fracture path and by increasing grain size. A dramatic demonstration of the latter effect is shown in the work of Minakawa, Levan and McEvily[23] who tested an aluminium alloy with an unusually small grain size, less than 1 µm. The material showed no measurable closure at all. The lack of roughness-induced closure is attributable to the fine grain size, which produced a very flat fracture surface. The implication, however, is that other types of closure, such as plasticity- or oxide-induced closure, do not operate in this class of alloys at all. The resulting threshold value was relatively low, and did not change with R ratio.

Short cracks and notches

In many materials, cracks less than 1 mm in length tend to show faster growth rates and lower thresholds than long cracks. One of a number of reasons for this is closure: short cracks tend to show less closure, which builds up as the crack grows. Plasticity-induced closure develops as the crack wake increases, and roughness-induced closure increases as more microstructural features are traversed. Thus the behaviour of a short crack depends on its plastic zone size and on its grain size (or other microstructural parameter size) as far as closure is concerned.

For the same reason, cracks which develop from notches will initially show very little closure. This is one reason why cracks may initiate at notches and then become non-propagating after a certain amount of growth.

Environment

Environment can affect closure levels in a number of ways. Oxidizing environments, and others which generate bulky surface deposits, will encourage closure, as will environments containing debris or viscous fluids capable of entering the crack. Environment can also affect slip character, and thereby alter the degree of roughness-induced closure.

Temperature

Changes in temperature will affect the rate of reactions with the environment, and thus may alter effects such as oxide-induced closure mentioned above. Additionally, changes in temperature may affect the crack path by, for example, embrittlement effects, which will have repercussions for the degree of roughness-induced closure. Temperature changes, therefore, may be either beneficial or detrimental.

An 'intrinsic' threshold

The term 'intrinsic' or 'closure-free' threshold is used to indicate the value of ΔK_{eff} at the threshold. It has been shown that, for some materials, the value of this parameter is virtually constant to changes in R ratio, microstructure, etc., and this has led some workers to conclude that there is an intrinsic threshold value which is constant over a whole class of materials. For example, most steels demonstrate a closure-free threshold of about 3 MPa $\sqrt{m}$ [24] which coincides with the applied ΔK_{th} values of short cracks and cracks at reasonably high R ratios.

While this assumption of an invariant threshold value does not hold true in all cases, for instance under very high R ratios or certain environmental changes, it is certainly a useful starting point for theoretical models, and may provide a useful general rule in design and materials selection.

Concluding remarks

There is no doubt that crack closure is a highly influential mechanism which has changed our thinking about near-threshold crack propagation. Currently, however, there is a tendency to explain *all* threshold behaviour in terms of closure, in cases where other mechanisms may be responsible. For example, oxide-induced closure may only be important in certain steels, and recent evidence suggests that even in these alloys it may only occur during relatively high frequency loading (see Chapter 6). However, the volume of research work which has been generated on oxide-induced closure is very large, and it is often mentioned as a posssible mechanism in alloy systems which have never been shown to demonstrate it.

As a very general rule of thumb, one can say that closure is of vital importance in understanding R ratio effects and the effects of microstructure, and that it makes important contributions to the short-crack effect. In the case of other effects, such as environment and temperature, closure should be considered as only one of several operating mechanisms, any one of which may dominate in a given situation.

References

1. Elber, W. (1971) in ASTM STP486, *The American Society for Testing and Materials,* p. 230
2. Elber, W. (1970) *Engineering Fracture Mechanics,* **2**, 37
3. Blom, A. F. and Holm, D. K. (1985) *Engineering Fracture Mechanics,* **22**, 997
4. Ritchie, R. O., Yu, W., Blom, A. F. and Holm, D. K. (1987) *Fatigue and Fracture of Engineering Materials and Structures,* **10**, 343
5. Suresh, S. and Ritchie, R. O. (1982) *Metal Science,* **16**, 529
6. Suresh, S. and Ritchie, R. O. (1983) *Engineering Fracture Mechanics,* **18**, 785
7. Ritchie, R. O., Suresh, S. and Liaw, P. K. (1981) *First Ultrasonic Fatigue Conference* (Champion, PA, USA), AIME, USA, p. 443
8. Suresh, S. (1984) Report MRL E-153, Brown University, USA
9. Minakawa, K. and McEvily, A. J. (1981) *Scripta Metallurgica,* **15**, 937
10. Park, D. H. and Fine, M. E. (1984) In *Fatigue Crack Growth Threshold Concepts,* TMS-AIME, USA, p. 145
11. Morris, W. L., James, M. R. and Buck, O. (1983) *Engineering Fracture Mechanics,* **18**, p. 871
12. Petit, J. (1984) In *Fatigue Crack Growth Threshold Concepts,* TMS-AIME, USA, p. 3
13. Beevers, C. J. (1982) In *Fatigue Thresholds,* EMAS, Warley, UK, p. 17
14. McEvily, A. J. and Minakawa, K. (1984) In *Fatigue Crack Growth Threshold Concepts,* TMS-AIME, USA, p. 517
15. Gerberich, W. W. (1988) In *Fatigue 87,* EMAS, Warley, UK, p. 1757
16. Gregory, J. K. (1988) In *Fatigue 87,* EMAS, Warley, UK, p. 303
17. Fleck, Z. (1984) Report No. CUED/C-MATS/TR.104, Cambridge University Engineering Department, Cambridge
18. Suresh, S. and Ritchie, R. O. (1984) In *Fatigue Crack Growth Threshold Concepts,* TMS-AIME, USA, p. 227
19. Laugier, M. (1987) *Proceedings, IDFC5* (Cork, 1987)
20. Buck, O., Rehbein, D. K. and Thompson, R. B. (1987) *Engineering Fracture Mechanics,* **28**, p. 413
21. Lankford, J. and Davidson, D. L. (1984) In *Fatigue Crack Growth Threshold Concepts,* TMS-AIME, USA, p. 457
22. Morris, W. L. and James, M. R. (1984) In *Fatigue Crack Growth Threshold Concepts,* TMS-AIME, USA, p. 479
23. Minakawa, K., Levan, G. and McEvily, A. J. (1986) *Metallurgical Transactions,* **17A**, 1787
24. Kendall, J. M., James, M. N. and Knott, J. F. (1986) In *The Behaviour of Short Fatigue Cracks* (EGF1), MEP, Bury St Edmunds, UK, p. 241

4 Methods of measurement

Testing equipment for threshold measurement on a CT specimen using both optical and clip-gauge monitoring, (Williams *et al*. [21]

Introduction

This chapter reviews the various experimental methods used to measure ΔK_{th} and near-threshold crack growth rates, and discusses problems and inaccuracies which arise during measurement. The chapter begins with a brief survey of the basic methods and equipment used, and follows with details of the proposed British and ASTM Standards which are being developed in this area. Problems associated with various methods, especially load-reduction methods, are then discussed in detail.

A thorough understanding of the experimental techniques involved in threshold testing is important for all users of threshold data. Owing to the lack of standardization, quoted threshold values will be found to vary depending on the test technique and on the method of defining threshold. It is therefore necessary to ascertain how data were arrived at before attempting to apply them.

Testing machines

Most threshold testing is carried out on servo-hydraulic and related testing machines, which are now in common use for experimental fracture-mechanics work. Considerable use is also made of electromagnetic resonance-type machines, since they offer the possibility of high-frequency loading up to ultrasonic frequencies. However, it is also possible to obtain useful threshold data from very simple fatigue machines such as Wohler-type rotating bending machines. The more sophisticated machines offer the advantages of accurate load control, which can be vital in the near-threshold region where the growth-rate curve is so steep that minor fluctuations in applied stress can cause major changes in growth rate, and may give rise to stress-change effects such as retardation (see Chapter 7). The load levels involved in threshold testing are generally very small compared to those used, for example, in fracture toughness testing; this may be a problem when using high load capacity machines at a very small fraction of their load capability. Another considerable advantage of servo-controlled machines is the possibility of computer control; for example, a complex load-shedding procedure which requires gradual load-range reduction dependent on the data from crack length monitoring equipment, can be fully automated. Even at relatively high frequencies, a single threshold test can take several days or even weeks to perform, a procedure which is extremely tedious if manual operation is required.

Specimens

It is common to use standard fracture mechanics specimens, such as compact tension (CT) and single edge notched bend (SENB) specimens, which are employed in fracture toughness testing. Other, 'applications-related' specimens have also been designed in order to duplicate particular in-service conditions. These may involve, for example, surface cracks or cracks initiating from notches, as well as specimens designed to impose complex loading patterns such as mixed-mode loading.

Generally speaking, the low load levels normally employed should allow more latitude in the choice of specimen size without encountering the type of size effects which occur in fracture toughness testing. However, there are distinct differences between fatigue crack propagation rates in plane-stress and plane-strain loading; normal fatigue specimens tend to show bowed crack fronts in which the crack is shortest at the surfaces of the specimen, implying that plane-stress conditions give rise to slower propagation and therefore higher thresholds. The existence of a larger amount of closure in plane stress may account for this (see Chapter 3). Some studies are available on the effect of specimen geometry on threshold, but their conclusions are not entirely consistent. Considering specimen thickness, for example, two studies[1,2] showed no pronounced differences over the range 5-40 mm, but three others[3,4,5] showed higher values of measured threshold in thicker specimens. This effect opposes the plain-stress/plain-strain effect mentioned above; oxide-induced closure may be responsible[4]. Tokaji, Ando and Nague[5], who recorded a slight increase in threshold with crack length for moderately thick specimens (8-22 mm), measured higher thresholds for 2 mm specimens, which remained high even at high R ratio. Closure was advanced as the reason for this effect, but the persistence at high R values suggests a more fundamental difference between plane-stress and plane-strain conditions. Plain-stress effects may be minimized by the use of side-grooved specimens (e.g. [6]).

Considering specimen type, Blom (Aeronautical Research Institute of Sweden) has noted a problem concerning CT specimens; slight misalignment of the specimen in the testing machine can give rise to significant shear loading across the crack faces. This seems to cause higher values of threshold, for reasons which are discussed under 'mixed-mode loading' in Chapter 7. The implication is that more symmetrically loaded specimens, such as centre-cracked plates, may give more reliable results. Vecchio, Crompton and Hertzberg[7] noted a similar effect, in which more symmetric specimens gave lower threshold values than CT specimens; they advanced a different explanation, based on stress distributions, but misalignment may have been responsible. To examine this trend we can consider results on a popular class of materials, the quenched and tempered low-alloy steels, compiled from a variety of sources[8]. Since the variation of threshold with yield strength within this class is fairly uniform, it is possible to check for the effects of a third parameter, such as specimen type. In this case, 24 results were available for which specimen type was known, and it is possibly significant that all the tests which used centre-cracked plates gave below-average threshold values. Results from CT specimens spanned the whole range of values, which might be the case if some specimens had been well aligned and others misaligned.

Considering the effect of crack length, a number of points should be noted. It is well documented that short cracks exhibit anomalous near-threshold behaviour, and this subject is covered at length in Chapter 8. This generally only applies to sub millimetre cracks, however. When dealing with cracks of a few millimetres in length, relatively high loads may have to be used, especially in the early stages of a load-shedding procedure. Such loads may approach yield-point magnitude; in such cases the guidelines used for K_{IC} testing may usefully be applied to check for the existance of yielding effects. Finally, some workers have suggested that crack length and load-shedding procedure may exert effects, even for long cracks; this will be discussed below.

Measurement of crack length

The accuracy of near-threshold growth-rate data and the total time involved in obtaining threshold results, are greatly influenced by the accuracy and stability of the crack length monitoring apparatus. Four commonly used methods will be compared below; other methods include systems derived from NDE equipment such as acoustic and ultrasonic systems.

1. Electrical potential drop

The growth of a crack increases the electrical resistance of the piece of metal, which can be detected as an increase in the potential drop in the region of the crack subjected to a constant applied current. Either AC or DC current can be used; Figure 4.1 shows a typical arrangement for a DC

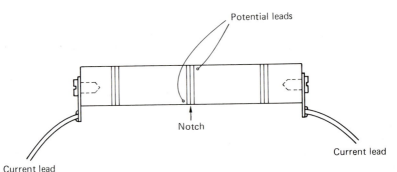

Figure 4.1 Typical arrangement of current and potential leads on a single-edge notched bend test specimen

potential drop system on an SENB specimen; Figure 4.2 shows a calibration curve for the same specimen. Typically, a constant current of 20–30 amps is used, requiring a power supply with good long-term stability. Typically, a steel specimen of the geometry shown in Figure 4.1, with a thickness and width of 20 mm and a 5 mm starter notch, will give an initial potential drop of about 200 µV; crack growth of 5 mm will increase this value to about 400 µV. With suitable amplification a change of 1 µV should be easily detectable, giving an accuracy of about 20 µm in crack length. Accuracy can be improved by placing the probes closer to the notch root, but this introduces more non-linearity into the relationship between potential drop and crack length.

Measurement of the absolute value of crack length is limited by long-term drift in the apparatus, but the method can be very accurate for measuring *changes* in length, and therefore establishing growth rates. A

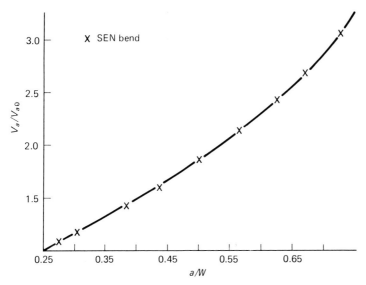

Figure 4.2 Potential drop calibration curve for SEN bend specimens. Crack length, a, is normalized by specimen width, W, and potential drop reading, V_a, is normalized by the value at zero crack length, V_{ao}

disadvantage is that the crack length value obtained is an average length for the whole crack front which may, as noted above, be bowed. In principle, the system can be calibrated for any specimen geometry. The AC systems tend to be capable of higher detection accuracy, especially for shorter cracks, but subject to more calibration difficulties.

2. Compliance

The elastic compliance of a specimen, measured as the ratio between any elastic deflection and the applied load, tends to increase with crack length. Compliance can be successfully used both as an indication of crack initiation and as a measure of crack length. Measurement of loading-point deflection is possible, but tends to be subject to noise and low sensitivity, so it is usual to measure deflection through a crack-mouth gauge or strain gauge attached to the specimen. These systems are covered in Chapter 3, since this is also a method of measuring crack closure.

If gauge output is carefully filtered, and if accuracy is improved using a derivative process (see Chapter 3), this method can be made as accurate as the potential drop method. On the whole, however, the potential drop method is probably easier to use. The compliance method has the advantage that it may be usable in circumstances where the electrical method is either unstable or liable to affect the results; examples include testing in corrosive solutions and at elevated temperatures. In the author's experience, a DC potential drop system was found satisfactory for measuring crack growth in a steel tested in sodium chloride solution, but this had to be substituted for a compliance system when a copper alloy was tested in a similar environment[9]. High-temperature testing can be a problem for the electrical method owing to thermocouple effects, but modern methods of signal processing can eliminate this problem.

Crack growth which involves a large amount of branching, deflection or closure may also lower the accuracy of these monitoring systems[10].

3. Optical methods

Direct observation of the crack, using a microscope or by taking period replicas of the surface, is still the most sensitive method of measuring crack growth. Even with very simple equipment one can detect changes in crack length of less than $10\,\mu$m, and the replication method, using acetate tape, allows a permanent record of the crack length at various times during the test. The method is, however, extremely tedious and time-consuming for the experimenter, and has two technical disadvantages: first, it only provides information about the *surface* growth of the crack, which may be quite different from its behaviour in the centre of the specimen; and second, it is necessary to interrupt the cyclic loading in order to make a reading (except for synchronized photographic methods); this 'dwell' period may affect subsequent crack growth.

The replication method is a very reliable means of studying short-crack growth and of viewing interactions between the growing crack and microstructure[11]. A second replication process can be used to obtain a metal replica which can be viewed in a scanning electron-microscope[12]; such replicas are capable of reproducing very fine details of the crack tip and surrounding microstructure.

4. 'Crack gauge' system

A gauge is used which resembles a strain gauge and can be attached to the surface of a specimen. As a crack grows through the metal beneath, it breaks successive elements of the gauge, altering the electrical signal from it. The accuracy of the gauge depends on the number of parallel wires which can be attached and on the adhesion to be metal. Surprisingly good accuracy is possible, and the system is convenient for automation and for use in unusual environments.

Methods of threshold measurement

The techniques used for obtaining ΔK_{th} values and/or near-threshold growth rate values can be divided into four categories:

1. Load-controlled K reduction

Most experimenters use a system in which a crack is grown in a standard fracture-mechanics specimen, such as a CT specimen; after growing the crack to a reasonable length, the applied ΔK is gradually reduced until the threshold is reached. Gradual reduction is necessary because a rapid decrease in ΔK gives rise to a phenomenon known as 'retardation'. This effect, which is discussed in Chapter 7, leads to unusually low crack growth rates and high thresholds. The implications of this are discussed in detail below.

It is usual to monitor crack growth rate continuously, reducing ΔK until the growth rate drops below some specified value used to define the threshold. Then ΔK may be gradually increased to provide further data points for the growth-rate curve. The national standards for threshold testing, which are discussed below, all use variations of this load-shedding method.

2. Strain-controlled K reduction

Gradual load shedding can also be achieved through control of some strain parameter such as crack-mouth opening or back-face strain (measured through a gauge mounted on the face opposite to the notch). With appropriate specimen geometry, a suitable rate of load shedding can be achieved simply by maintaining a constant value of the chosen strain range [13], making the automation of load shedding somewhat easier, but introducing some difficulties which will be discussed below.

3. Klesnil and Lucas method

This method [14] involves the deliberate application of a sudden decrease in ΔK, which causes retardation. This enables the experimenter to measure, relatively quickly, a 'pseudo threshold', this being the value of ΔK_{th} immediately following the sudden load reduction. This procedure is repeated using successively smaller values for the load reduction, giving decreasing values of the pseudo threshold. It is argued that the 'true' threshold will correspond to the value of the pseudo threshold when the load drop reaches zero, and that this point can be found by extrapolation.

The method is not recommended owing to the difficulties and assumptions inherent in this extrapolation procedure.

4. Stress-relief methods

In some materials it is possible to circumvent the problem of retardation by carrying out a stress-relief annealing treatment after load reduction. The procedure is thus: (a) crack growth is initiated at a relative ΔK and the crack is grown to a suitable length; (b) the specimen is removed from the testing machine and stress-relieved by annealing; (c) testing is continued at a low applied ΔK, close to the expected threshold value.

Essentially a large, stepwise reduction in ΔK is achieved, using an intervening annealing treatment. This method has the obvious advantages that it saves both time and crack length. For various reasons, the initial growth which occurs from the stress-relieved crack at the beginning of stage (c) above may be uncharacteristic (see Chapter 9), so it is advised that the crack be allowed to grow for a short distance before results are recorded.

Pook and co-workers [15] have used a simple version of this method for many years, enabling threshold values to be obtained from testing machines which are normally used for conventional S/N-type tests. In this case the specimen is first cycled at a high load in order to initiate a crack, whose dimensions are not known. Having produced a number of such pre-cracked specimens, they are then tested at various applied stress ranges. Only after each specimen has failed is the value of ΔK for crack growth from the pre-crack calculated, based on the observed pre-crack size and shape. Data are recorded as a type of S/N curve, with ΔK replacing stress on the vertical axis (Figure 4.3). The 'fatigue limit' on this curve corresponds to the threshold.

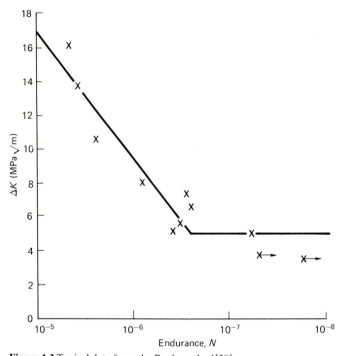

Figure 4.3 Typical data from the Pook method[29]

The chief advantage of this method is that it can be used on very simple test equipment, using simple testing routines. Equipment can be easily adapted to perform tests in, for example, corrosive solutions or elevated temperatures. It is, however, expensive on specimens, since it involves a 'hit-or-miss' procedure in which many specimens may be wasted before useful data are obtained.

Another method, developed by the author[16], combines the standard load-shedding procedures with stress-relief annealing. In this case, normal crack monitoring methods are used, but a period of stress relief is employed to cut out the long initial process of load reduction. It has been shown for a copper alloy that this method gives identical near-threshold data to that obtained using more conventional load shedding[16].

A problem associated with both the above methods is that many materials cannot be fully stress-relieved without changing their microstructure and properties. This would be the case for many high-strength steels and aluminium alloys in their quenched-and-tempered conditions, for instance.

National standards for threshold testing

Both the British and American Standards authorities are developing standards for crack propagation testing in the near-threshold regime. In addition there is already an American Standard for testing at the higher crack growth rates, above 10^{-5} mm/cycle (ASTM E647-83).

British Standard

A draft for development (DD) for a British Standard has been produced, entitled *Draft British Standard Method for the Measurement of Threshold Stress Intensity Factors and Fatigue Crack Growth Rates in Metallic Materials*. This is being circulated to all interested parties and comments invited, through either the British Standards Institution or Dr Trevor Lindley, CERL, Leatherhead, UK.

The DD recommends two alternative procedures, both of which involve load shedding under load control, after pre-cracking in a standard fracture-mechanics specimen. The principal method involves continuous reduction of K at constant R, keeping constant the value of C in the equation:

$$C = \frac{1}{\Delta K} \frac{d(\Delta K)}{da}$$
(4.1)

This produces an exponential decrease of ΔK with crack length (see Figures 4.5 and 4.6 below). The specified value of C is $-100 \, \text{m}^{-1}$ (i.e. $-0.1 \, \text{mm}^{-1}$), which gives a relatively rapid load-shedding rate. The method involves an initial estimate of the expected value of ΔK_{th}, followed by checks on the validity of the procedure once this provisional value for ΔK_{th} has been measured; this resembles the approach in the BS for fracture toughness testing.

The second method proposed is a reduction of ΔK at constant K_{max}. This method, the merits of which will be discussed below, involves decreasing ΔK in a stepwise fashion by increasing K_{min}. Inevitably, the measured threshold value occurs at a relatively high R ratio.

The threshold is defined as being reached if no crack growth can be detected for a period $q/10^{-11}$ cycles, where q is the resolution of the crack monitoring equipment, in metres. A resolution of better than 0.1 mm is required. This definition corresponds to a da/dN value of 10^{-8} mm/cycle, which is the same as that recommended by the author (see below).

ASTM Standard

Load shedding is required, either continuously using constant C parameter as above, with $C = -80 \, \text{m}^{-1}$, or in incremental steps in which the load is reduced by no more than 10%, after at least 0.5 mm of crack growth per step. This would be equivalent to continuous reduction at a C value of about $-200\,m^{-1}$, i.e. a more rapid rate than that recommended for continuous load shedding. In the author's experience, reduction in steps of 10% is satisfactory at high stress intensities, but once near-threshold conditions are reached, such a reduction may cause retardation of growth (see Chapter 4), leading to erroneously high estimates for ΔK_{th}.

The ASTM method of determining ΔK_{th} from growth-rate data also differs from the BS recommendation. The threshold is defined as the value of ΔK at which $da/dN = 10^{-7}$ mm/cycle. This value is to be established by linear extrapolation of data in the region 10^{-7} to 10^{-6}mm/cycle, using at least five data points.

There are two problems associated with these definitions: firstly, the value of ΔK at 10^{-7}mm/cycle may be significantly higher than the 'true' threshold, i.e. the value of ΔK_{th} at which no growth occurs. This problem is discussed below, under 'definition of threshold'; second, it is often difficult to establish as many as five distinct data points at these low crack growth rates. A number of workers have reported difficulties in meeting this requirement; using the stepwise procedure, the establishment of five data points in this region at a test frequency of 10Hz would take almost two weeks of continuous cycling, so low-frequency testing is effectively impossible.

Load-shedding methods: the effect of various testing parameters

It is now appropriate to examine the various load-shedding methods in more detail, showing how results can be affected by various aspects of the test technique.

A number of workers have suggested that the results obtained from such tests are dominated by stress-history effects, and thus cast doubt on the great majority of available threshold data. It is argued that the residual stresses which cause closure continue to act in the crack wake, even at large distances behind the crack tip, so that crack arrest is controlled by the load-shedding sequence. Typical of experiments in this area is the work of James and Knott[17]. Using a normal load-shedding sequence, they grew a crack to near-threshold growth rates (Figure 4.4) before machining away the crack wake up to 0.5 mm from the crack tip. On returning the specimen to the machine they found an increase in da/dN (the arrow to open circle on Figure 4.4) and a decrease in measured closure levels. From this they concluded that closure stresses acting more than 0.5 mm behind the crack tip, had been operating before the wake was removed.

These observations certainly give cause for concern, but two other explanations may be suggested. First, the test was conducted at a relatively high R ratio, 0.35, so that the act of *unloading* the specimen from the testing machine would of itself tend to induce a growth transient (see section on underloads in Chapter 7). Second, the removal of material, which was done by electro-discharge machining, can give rise to significant residual stresses.

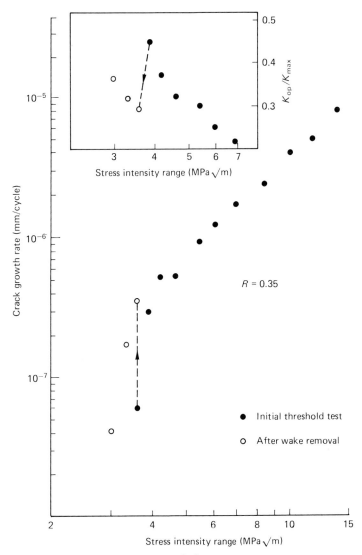

Figure 4.4 Results of James and Knott[17] demonstrating changes in crack growth rate and closure after removal of the crack wake

Despite some reports of this type, the great majority of work available to the author suggests that, provided general guidelines are adhered to, there is a surprising degree of consistency between the results of different workers using load-shedding techniques; this would not be expected if the above observations were generally valid.

Figure 4.5 shows schematically the three types of gradual load reduction which can be employed; a continuous decrease of ΔK with increasing crack length is preferred, though this can really only be achieved using either digital control methods or strain control. In practice, a reasonable

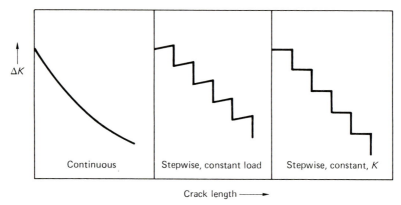

Figure 4.5 Alternative methods of load reduction

compromise is to decrease the load in small steps. Within each step, constant load or constant K conditions can be maintained. In any case, the rate of decrease of ΔK is critical and must be carefully controlled.

Continuous K-decreasing

The decrease of ΔK according to the C parameter mentioned above, where:

$$C = \frac{1}{\Delta K} \frac{\mathrm{d}(\Delta K)}{\mathrm{d}a} \tag{4.2}$$

was proposed some years ago by Saxena *et al.*[10]. If C is kept constant, ΔK decreases exponentially with crack length. This is illustrated in Figure 4.6 for a C value of $-0.16\,\mathrm{mm}^{-1}$. C values are always negative; numerically larger values correspond to more rapid rates of reduction. Under this type of load shedding, a given amount of crack advance is accompanied by a decrease in ΔK by a constant *proportion*; this also therefore causes a decrease in the plastic zone size by a constant proportion.

Table 4.1 gives information from various workers who have used this C parameter approach[1,2,10,18–21]. In most cases, consistent results were obtained for all the C values used. The results of Bathias[1] were obtained from 'round robin' tests conducted by three laboratories using the same stainless steel, with a stepwise load shedding procedure which corresponded to a C value of about $-0.33\,\mathrm{mm}^{-1}$. Considerable scatter was displayed, with threshold values varying from 3.7 to 7.0 MPa $\sqrt{\mathrm{m}}$, suggesting that this reduction rate is too high for this material.

Hudak *et al.*[18] varied the value of C systematically, with surprising results; at high R ratio there was little change to the measured value of ΔK_{th}, but at $R = 0.1$ the measured threshold was relatively high for the low reduction rates ($C = -0.02$ to -0.09) and lower for the highest reduction rates, up to $C = -0.3$. A similar effect was demonstrated by Cadman, Brook and Nicholson[19]. This is contrary to what might be expected; high reduction rates should cause anomalously high values for ΔK_{th} owing to

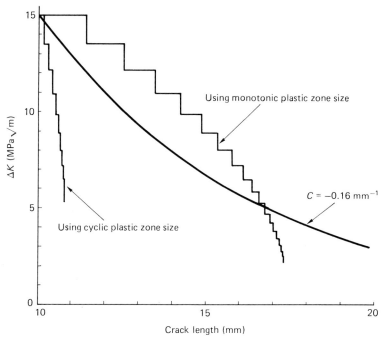

Figure 4.6 Typical load shedding, comparing reduction rates using the constant-C method and the plastic zone size method

retardation. In fact, the complete dependence of ΔK_{th} on reduction rate has been shown to demonstrate a minimum, as shown in Figure 4.7, due to Brook[22]. The data in this diagram show a particularly large variation; presumably the effect is not so great for most materials, since workers using a variety of reduction rates are generally able to obtain similar results. However, such effects do raise important questions as to the nature of the threshold parameter.

Table 4.1 The The C parameter approach

Author	C values used (mm^{-1})	Comments
Hudak et al.[18]	-0.02 to -0.3	See text
Cadman et al.[19]	-0.08 to -0.05	See text
Saxena et al.[10]	-0.024 and -0.04	Aluminium alloy
Hertzberg et al.[20]	-0.059	Two aluminium alloys
Saxena et al.[10]	-0.069	Steel
Williams et al.[21]	-0.098	Copper
Mingda et al.[2]	-0.1	Little data at the low da/dN values. Same threshold obtained by a stepwise decrease method
Bathias[1]	-0.33	See text

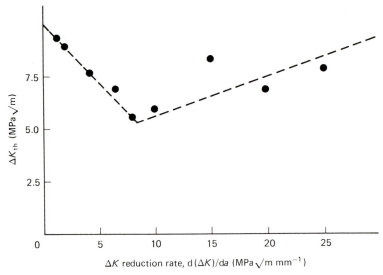

Figure 4.7 Variation of measured threshold with load reduction rate, from Brook[22]

In this case the increase in measured threshold at very low reduction rates seems to be due to increased closure, probably caused by oxide and other debris, arising from the growth of the crack for long periods at near-threshold stress intensities. Such test conditions may be appropriate for certain applications, as will be discussed below, but for general use the aim must be to find the lowest possible value of ΔK at which propagation may occur, as only thus can we be sure of a conservative figure.

In the light of the above remarks, the C values chosen by the ASTM and British draft Standards, of -0.08 and $-0.1\,\mathrm{mm}^{-1}$ seem to be on the high side, but not excessively so. The alternative, stepwise procedure in the ASTM version, for which $C = -0.2$, may be too high for safety. In any case, it is recommended that, for any large testing programme on a new material, a range of reduction rates should be investigated at the outset in order to avoid some of the problems listed above.

Load shedding controlled by plastic zone size

This alternative load-shedding method has been used by many workers in the past; indeed the majority of threshold data now available has probably been obtained in this way. The method involves stepwise reduction of ΔK in which the amount of crack growth during each step is related to the plastic zone size corresponding to the previous step. The rationale for this is that, after ΔK is reduced, the crack must initially grow through the plastic zone which was set up during the previous cycling. While it passes through this plastic zone, retardation effects are likely, since the plastic wake immediately behind the crack contains residual stresses and also possibly because the material in the plastic zone has been cyclically hardened.

If the crack growth rate is being continuously monitored, the end of this retardation period can be detected. However, most workers choose to grow the crack for a distance equal to several times the plastic zone size; a factor of four is commonly used. There is some variation between researchers in terms of the method of calculation of plastic zone, and of the choice between cyclic and monotonic plastic zone sizes. Typical load-shedding sequences are shown in Figure 4.6 compared with a typical C-parameter sequence. A material yield strength of 450 MPa has been assumed here – the plastic zone size will of course vary strongly with yield strength. The difference between the two methods is clear; when plotted in terms of crack length, the plastic zone method shows an increasing rate of ΔK-reduction as ΔK_{th} is approached; the opposite effect is shown by the C-parameter method. While the rationale behind the plastic zone approach seems reasonable with regard to the immediate environment of the crack-tip, a number of workers have voiced the fear that crack-tip conditions, and especially closure levels, may be influenced by conditions in the wake of the crack at large distances from the crack tip.

Considering the plot of Figure 4.6, and assuming a threshold at 8 MPa $\sqrt{m}$, it can be seen that, if cyclic plastic zone size is used, the total amount of crack growth employed will be only 0.65 mm, which is about twice the size of the *monotonic* plastic zone size for the *initial* ΔK value (15 MPa $\sqrt{m}$). Under such circumstances we cannot be sure that we are not building up stress-history effects from previous load levels.

No systematic studies are available which investigate the plastic zone size method. Bathias *et al.* [21], who conducted the 'round robin' tests mentioned above, also tested an aluminium alloy using a plastic zone method in which the growth increment was fixed at four times the reversed plastic zone size; consistent results were achieved. The author, in a study which is reported at the end of this chapter, has compared this method with others, finding no significant differences.

One advantage of the plastic zone method is that threshold conditions can be achieved at much shorter crack lengths; the C-parameter method must generally be used on wide plates or CT specimens, whereas the plastic zone method can be used on a variety of specimen geometries.

Load shedding using Q parameter

Bailon, Chappuis and Masounave [23] have suggested a more rapid method of obtaining threshold values, based on a parameter, Q, where:

$$Q = \frac{1}{\Delta K} \frac{d(\Delta K)}{dN} \tag{4.3}$$

This parameter resembles C, except that number of cycles, N, is substituted for crack length. Use of constant Q during load shedding leads to a downward curving plot, as shown in Figure 4.8. The procedure involves starting the load shedding at a relatively high value of Q; the value is then reduced in stages, between which are periods in which ΔK is allowed to rise, as shown in Figure 4.7. The final stages of this procedure, in which the threshold is reached by successively lower Q rates, is not unlike the approach of Klesnil and Lucas mentioned above. For the data

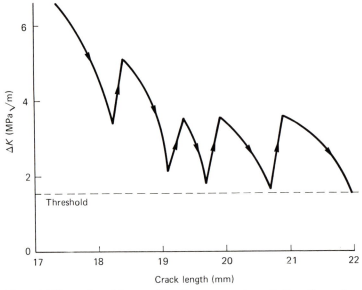

Figure 4.8 Illustration of the use of the Q parameter, due to Bailon, Choppuis and Masounave[23]

reported here, even the final Q value corresponded to quite a high C value, of about $-0.5\,\mathrm{mm}^{-1}$.

Load shedding at constant K_{max}

This method, proposed by Doker, Bachmann and Marci[24] and since investigated by others[25,26], involves load shedding while keeping K_{max} constant, thus reducing ΔK by increasing K_{min}. The advantage of this procedure is that relatively large rates of reduction are possible without causing retardation, though it has been shown that *very* large reduction rates lead to lower values of measured threshold. Using this method, it is possible to arrive at the threshold in a relatively short space of time. The major disadvantage, of course, is that the R ratio is constantly increasing, so this is only an appropriate method for high R values.

The method is recommended in the British Standard draft for threshold testing discussed above, but not in the ASTM version.

Load shedding under strain control

Deans and Richards[13] developed a method in which the controlling parameter was back-face strain, i.e. strain measured from a gauge fastened on that face of the specimen which is opposite to the initiating notch or crack. Since, as crack growth proceeds, the specimen becomes more compliant, if the range of strain is kept constant, this will involve a gradual reduction in load levels. Blom, Hadreboletz and Weiss[27] used both back-face strain and crack-mouth monitoring; for their material and

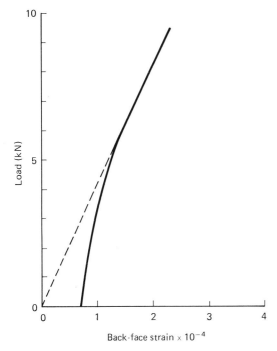

Back-face strain × 10^{-4}

Figure 4.9 Experimental variation of back-face strain with load, from Mayes and Baker[28]

specimens the rate of load shedding corresponded approximately to $C =$ $-0.08\,\text{mm}^{-1}$, i.e. a rate which is similar to that generally employed in continous load-controlled methods.

Mayes and Baker[28] identified a problem associated with this method, which is illustrated in Figure 4.9. The relationship between applied load and strain, which should be linear, becomes non-linear at low load values. This is presumably caused by closure (see Chapter 3), and is more pronounced as crack length increases. The effect of this is that, as the test proceeds, R tends to decrease. There is also a 'residual' strain, i.e. a positive value at zero load. To avoid this problem, Blom, Hadrboletz and Weiss[27] arranged for the back-face strain to be continually adjusted during the test, so as to keep R constant. This, however, tends to negate one of the advantages of this method, which is the relative ease of automatic control that it offers.

An advantage of the method is that it also enables one to measure closure values (see Chapter 3) and therefore to calculate effective, closure-free stress intensity values.

The definition of threshold

In principle, the threshold value corresponds to the condition of zero crack growth, i.e. to the asymptote of the growth-rate curve. In practice, most data do not show a clear asymptote even at the lowest growth rates that can be measured. A more practical definition is the value of stress intensity range corresponding to a specific growth rate, chosen to be so low that crack growth will be negligable for all practical situations. This definition is analogous to that of the 'endurance limit' defined for S/N data which do not show a clear 'fatigue limit'.

The ASTM Standard for threshold testing defines the threshold at a growth rate of 10^{-7} mm/cycle, whereas the draft British standard uses a rate of 10^{-8} mm/cycle or lower. The author's opinion is that 10^{-7} mm/cycle is too high a value, for two reasons. First, it corresponds to a significant amount of crack growth for many typical high-cycle applications. Components in, for example, the transport and machine tool industries, are often required to withstand 10^8 cycles, therefore the endurance limit stress is often chosen at this value. At a growth rate of 10^{-7} mm/cycle, this number of cycles would allow 10 mm of crack growth, which would be sufficient to cause failure in many components, even if the inevitable acceleration of growth rate is not considered.

Table 4.2 Percentage difference between ΔK_{th} and ΔK at $da/dN = 10^{-7}$ mm/cycle

	(%)
Copper	25
Mild steel	20
13% Cr steel	14
Aluminium/magnesium alloy	20
Copper in NaCl solution	38
Stainless steel in NaCl	28
Mild steel in NaCl	38

The second reason for preferring a lower growth rate for definition purposes is that there are significant differences between behaviour at 10^{-7} mm/cycle and behaviour at lower growth rates. This is dramatically illustrated by data obtained at ultrasonic frequencies which enable one to measure much lower growth rates than conventional test methods, often including 'true' asymptotes. Table 4.2 presents some data from this type of testing[8], showing the percentage difference between the true threshold and the value of ΔK at 10^{-7} mm/cycle. It is important to note that not only are the differences quite large, but they vary according to both material type and environment. Thus even the qualitative comparison of two materials or two environments may not be the same at 10^{-7} mm/cycle as it is at lower growth rates.

For the above reasons, a threshold defined at 10^{-8} mm/cycle is to be preferred. For the great majority of data, this will fall within 10% of the true asymptote, and in any case it corresponds to a growth rate sufficiently low as to be negligible for most applications.

In compiling a compendium of fatigue threshold values[8], the author used the following guidelines for selecting data and for defining ΔK_{th}:

1. Some data points should be available at growth rates below 10^{-7} mm/cycle. Growth-rate curves which were measured entirely above this value, even if they seemed to show an asymptote, were rejected as unreliable.
2. The value of ΔK_{th} was defined at 10^{-8} mm/cycle, using a best-fit line to all data available at growth rates below 10^{-6} mm/cycle.

Scatter in threshold results

Though there have been relatively few studies undertaken to establish the degree of variability of measured threshold values, it is possible to get an impression of this by examining results from commonly tested classes of materials such as the alloy steels and certain commercial aluminium alloys. It seems that the value of ΔK_{th}, while not as precisely measurable as, for instance, a proof stress, nevertheless shows less inherent scatter than, say, a fatigue limit value.

While some workers have recorded worrying variations resulting from changes to test procedure, the experience of the majority is that, provided one uses a standard load-shedding procedure with reasonable care, a reproducible threshold value should be obtained with a scatter of no more than 10% (i.e. plus-or-minus 5% from a mean value). If one compares results from different test methods, and from different batches of material, a further 10% scatter may appear, giving a total of 20%. This second 10% will probably be eliminated with improved understanding of test methods and of the effects of material parameters.

From the designer's point of view, assuming that only one value of ΔK_{th} for the material in question is available, a suitably conservative approximation would be to reduce this value by 20%, though this will not, of course, allow for any real differences which may exist between the testing situation and the application, such as variable amplitude loading or environmental effects.

Other measurement problems

This section considers a number of problem situations in which the measured threshold may differ from that which operates in practice. Needless to say, one should always attempt to reproduce in laboratory tests the same conditions as are expected in service, including material condition, environment and loading pattern, but in some cases this will not be feasible. For a detailed discussion of particular effects, the reader is referred to the relevant chapter in this book.

'Pseudo thresholds' in aqueous solutions

Some tests conducted in corrosive, aqueous solutions may yield apparent asymptotic behaviour at relatively high da/dN values, which turns out not to be the threshold, but a temporary step in the growth-rate curve. This effect is outlined in Chapter 6 (see Figures 6.2 and 6.3), and has led a number of workers to report threshold values which are erroneously high. If the guidelines mentioned above are followed with regard to the minimum growth-rate values to be measured, this problem will be avoided.

Frequency effects

Frequency effects are a major problem in all high-cycle fatigue work; it is almost impossible to conduct meaningful threshold tests at frequencies below 10 Hz, but many applications involve much lower frequencies and yet still require high-cycle lifetimes. Examples are offshore structures subjected to wave loading, and artificial hip joints; the latter are loaded at walking frequency, about 1 Hz, and are required to last for up to 50 years at an estimated 2 million cycles per year.

In Chapter 6, two types of frequency effect are discussed: first, most corrosive solutions give rise to some form of frequency dependence of crack growth rate, which may be of a complex form involving a maximum point; second, recent evidence suggests that oxide-induced crack closure, an important threshold-increasing mechanism, may only operate at high frequencies, over 10 Hz. In both cases the consequence is that much of the available test data may be non-conservative with respect to da/dN values in the threshold region and to ΔK_{th} values themselves.

One possible method for obtaining some information at low frequencies is to approach near-threshold conditions by load shedding at a *high* frequency before reducing the frequency and observing any changes in growth rate. Assuming a detection accuracy of 10 μm crack growth, a growth rate of 10^{-7} mm/cycle can be detected after 10^5 cycles, which is about one day at 1 Hz. There is always a danger of transient effects occurring when the frequency is changed, and also of the specimen being in some way pre-conditioned by the loading history, but some useful indications of near-threshold frequency effects may be obtained in this way.

The choice of test method

From the above discussion, it is clear that there are a variety of test methods, all of which possess disadvantages, including those now recommended by standards authorities. It has been shown that, in a number of cases, the value of threshold obtained is conditioned by the choice of test method and by the choice of numerical parameters within that method. The reader may be forgiven for asking whether any of these measured values is the 'true' threshold, and what practical significance this true threshold has when applied to a particular industrial problem.

However, despite a few disturbing reports, the general impression one obtains from looking at any large collection of data on thresholds (e.g. [8]) is that results on many materials are surprisingly consistent, despite the variety of methods used to obtain them.

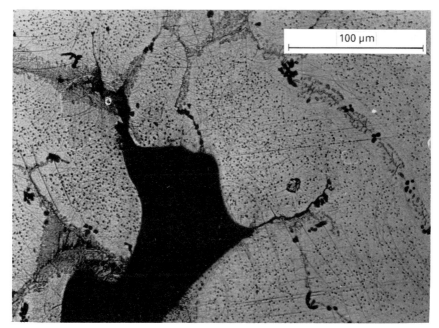

Figure 4.10 Typical shrinkage cavity in aluminium bronze casting used for comparative threshold testing

To illustrate this, the author has carried out a study on an aluminium bronze alloy[16]. In this case, threshold values were measured at various R ratios using four different methods:

1. Conventional load shedding using the plastic zone size method: here the growth increment was four times the reversed plastic zone size.
2. Load shedding at $R = 0.1$ using the same procedure, followed by a variation in R ratio achieved by increasing K_{min} at constant K_{max}: this enabled thresholds to be measured at various R ratios up to 0.7.
3. Load shedding using stress-relief heat treatment, as described above: in this case an annealing temperature of 300°C was sufficient to cause full stress relief without affecting microstructure. Pre-cracks were grown at high ΔK, followed by annealing, after which data were obtained immediately in the near-threshold regime.
4. Testing of natural defects in the form of shrinkage cavities: castings of the material contained small cavities whose re-entrant shapes gave them sharp, crack-like corners from which cracks grew easily (Figure 4.10). The casting had been very slowly cooled, giving effective stress relief. The defects involved were small (submillimetre) but not small enough to demonstrate the anomalous crack growth effects described in Chapter 8.

Figures 4.11–4.13 compare the four different methods. It may be seen that they all give comparable threshold results and near-threshold growth rates. For this material, at least, the threshold seems to be a material parameter which is unaffected by the method of measurement.

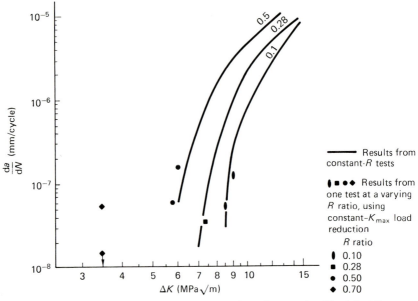

Figure 4.11 Thresholds in aluminium bronze; comparison of conventional load shedding (solid lines) and load shedding at constant K_{max} (data points), from Taylor[16]

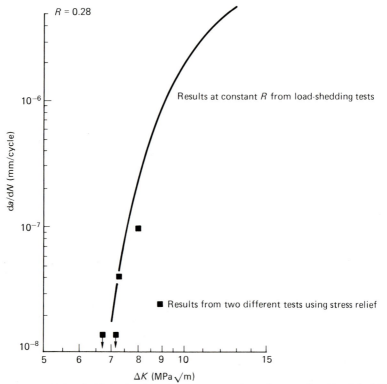

Figure 4.12 Thresholds in aluminium bronze; comparison of conventional load shedding (solid line) and load shedding assisted by stress relief (data points), from Taylor[16]

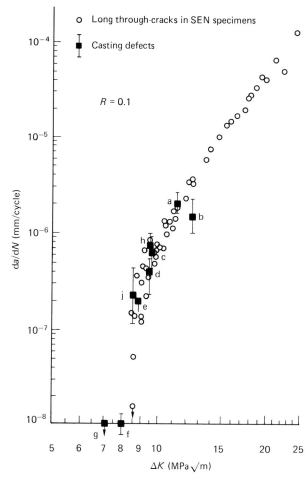

Figure 4.13 Thresholds in aluminium bronze; comparison of conventional load shedding (open circles) using long cracks, with data derived from naturally-occurring defects (e.g. Figure 4.10). from Taylor[16]

Application of threshold data

The use of threshold data in service situations is dealt with through a number of case studies in Chapter 10. However, it is appropriate here to say a word or two concerning the choice of test method for a given application.

Clearly, it is necessary to obtain threshold data for the same material condition and loading type as expected in the application, though small variations in, for example, R ratio or material yield strength, may be allowed for if data in the following chapters is consulted.

Some of the work reported above has suggested that stress and crack-length history also play a role. There is clearly a difference between a crack growing from a naturally occurring defect in a fully stress-relieved casting, say, and a crack growing from a machined notch which experiences a rapidly decreasing stress field as it grows. Assuming no large variations in load history, the majority of cracks in real structures will be not be subject to load-shedding sequences, but will be cycled close to their thresholds for long periods of time. Such cracks may be either short, and subject to relatively high loads, or long and subject to lower loads. The latter case corresponds more closely to a situation of very low C values, for which quite high thresholds have been measured (see above), whereas the shorter crack may be able to continue growing at low ΔK levels and will also tend to accelerate in growth owing to the high applied loads.

Clearly, the best test method is that which most closely approximates to the application. In this respect the problems of variable amplitude loading (Chapter 7) and defect type (Chapter 9) should be considered.

References

1. Bathias, C. (1981) *Fatigue of Engineering Materials and Structures,* **4**, 1
2. Mingda, G., Chuanfu, D., Wei, Z. and Minggao, Y. (1985) *Fatigue 84,* EMAS, Warley, UK, p. 287
3. Pook, L. P. and Greenan, A. F. (1974) NEL Report No. 571, The National Engineering Laboratory, UK
4. Romaniv, O. N., Tkach, A. N. and Lenets, Y. N. (1987) *Fatigue and Fracture of Engineering Materials and Structures,* **10**, 203
5. Tokaji, K., Ando, Z. and Nagae, K. (1987) Transactions of the American Society of Mechanical Engineering, *Journal of Engineering Materials and Technology,* **109**, 87
6. Venables, R. A., Hicks, M. A. and King, J. E. (1984) In *Fatigue Crack Growth Threshold Concepts,* TMS–AIME, USA, p. 341
7. Vecchio, R. S., Crompton, J. S. and Hertzberg, R. W. (1987) *Fatigue and Fracture of Engineering Materials and Structures,* **10**, 333
8. Taylor, D. (1985) *A Compendium of Fatigue Thresholds and Growth Rates,* EMAS, Warley, UK
9. Taylor, D. and Knott, J. F. (1978) *Metals Technology,* **10**, 221
10. Saxena, A., Hudak, S. J., Donald, J. K. and Schmidt, D. W. (1978) *Journal of Testing and Evaluation,* **6**, 167
11. Taylor, D. and Knott, J. F. (1981) *Fatigue of Engineering Materials and Structures,* **4**, 147
12. Brown, C. W. (1981) Unpublished PhD thesis, University of Cambridge
13. Deans, W. F. and Richards, C. E. (1979) *Journal of Testing and Evaluation,* **7**, 147
14. Klesnil, M. and Lucas, P. (1972) *Engineering Fracture Mechanics,* **4**, 77
15. Pook, L. P. (1972) ASTM STP 513, The American Society for Testing and Materials, p. 106
16. Taylor, D. (1989) *Engineering Fracture Mechanics,* **32**, 177
17. James, M. N. and Knott, J. F. (1985) *Fatigue of Engineering Materials and Structures,* **8**, 177
18. Hudak, S. J., Saxena, A., Bucci, R. J. and Malcolm, R. C. (1978) Report No. AFML/TR/78/40, Air Force Materials Laboratory, USA
19. Cadman, A. J., Brook, R., and Nicholson, C. E. (1981) In *Fatigue Thresholds,* EMAS, Warley, UK, p. 59
20. Hertzberg, R. W., Miller, G., Donald, K., Stofonak, R. J. and Jaccard, J. (1981) *Proceedings ICF5* (Cannes, France, 1981)
21. Williams, R. S., Liaw, P. K., Peck, M. G. and Leax, T. R. (1983) *Engineering Fracture Mechanics,* **18**, 953
22. Brook, R. (1983) In *Fatigue Crack Growth Threshold Concepts,* TMS-AIME, USA, p. 417

23. Bailon, J-P., Chappuis, P. and Masounave, J. (1981) In *Fatigue Thresholds,* EMAS, Warley, UK, p. 77
24. Doker, H., Bachmann, V. and Marci, G. (1981) In *Fatigue Thresholds,* EMAS, Warley, UK, p. 45
25. Herman, W. A., Hertzberg, R. W., Newton, C. H. and Jaccard, R. (1987) In *Fatigue 87,* EMAS, Warley, UK, p. 819
26. Castro, D. E., Marci, G. and Munz, D. (1987) *Fatigue and Fracture of Engineering Materials and Structures,* **10**, 305
27. Blom, A. F., Hadreboletz, A. and Weiss, B. (1983) *Proceedings ICM4* (Stockholm, 1983), p. 755
28. Mayes, I. C. and Baker, T. J. (1981) *Fatigue of Engineering Materials and Structures,* **4**, 79
29. Pook, L. P. (1975) *Journal of Strain Analysis,* **10**, 242

5 The effect of microstructure and mechanical properties

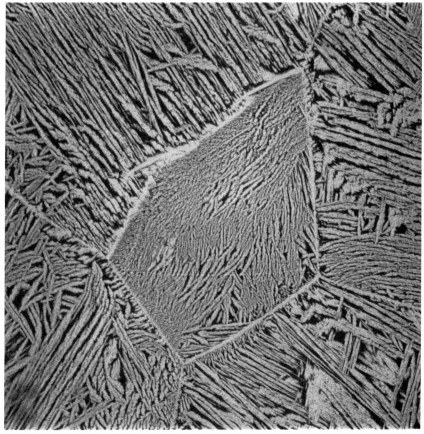

Microstructure of rapidly cooled aluminium bronze

Introduction

On an academic level, far more research has been conducted, and more papers written, on the effect of microstructure and mechanical properties than on any other topic covered in this book. Because of this, some property relationships have been clearly established in many different alloy systems; the effect of grain size on threshold is one example. However, a criticism of the work in this field is that it tends to be confined to a relatively small number of alloys, dictated by the requirements of certain high-technology industries. There is a wealth of data on certain alloy steels and 7000-series aluminium alloys, for instance, but very little information on threshold behaviour in, for example, cast irons, despite their widespread use and despite the fact that certain cast irons seem to have surprisingly high thresholds.

Material properties, both mechanical and microstructural, have a strong influence on near-threshold crack propagation. Unlike the Paris region, the near-threshold propagation curve is markedly affected by changes in yield strength, and in microstructural parameters such as grain size, as illustrated schematically in Figure 5.1. This is because the crack-tip deformations and crack-advance rates are of a scale commensurate with the microstructural scale of the material.

Figure 5.2 illustrates the importance of microstructural features on a crack growing at a near-threshold rate. Comparing the unetched and etched micrographs of this aluminium bronze alloy, one can see the effect of precipitates causing deflection and branching.

The difficulties associated with material properties such as yield strength and grain size are illustrated by some results on steels[1] shown in Figures

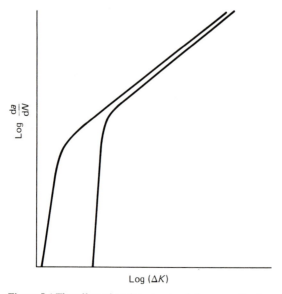

Figure 5.1 The effect of a microstructural change on the form of the growth-rate curve

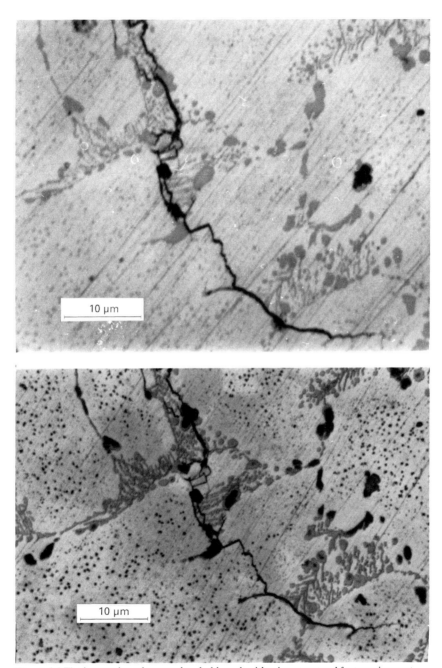

Figure 5.2 The interaction of a near-threshold crack with microstructural features in a cast aluminium bronze alloy; (*a*) unetched; (*b*) etched microstructure

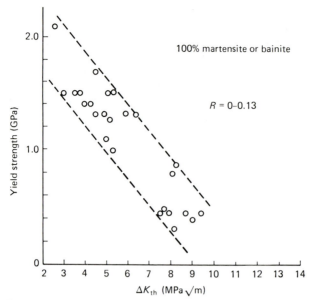

Figure 5.3 Variation of threshold with yield strength in a number of alloy steels, from Taylor[1]

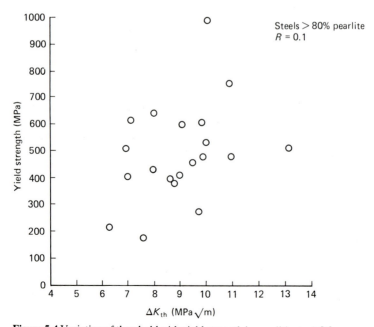

Figure 5.4 Variation of threshold with yield strength in pearlitic steels[1]

5.3–5.5. Figure 5.3 shows the variation of threshold with yield strength for a large number of alloy steels, all quenched and tempered to martensite or bainite. Despite the scatter, a consistent relationship emerges in which decreasing yield strength increases threshold. It will be shown below that it is in fact not yield strength but microstructure which is causing this change. Figure 5.4 shows a similar plot for fully pearlitic steels; in this case there is no clear relationship between yield strength and threshold. This is surprising since the microstructures of pearlite and bainite are, superficially, quite similar.

Figure 5.5 shows some data on low-carbon mild steels, for which a clear dependence on grain size appears; in fact the dependence on yield strength would be equally clear since the two are simply related through the Hall–Petch equation, provided the carbon content, and therefore the amount of pearlite, is low.

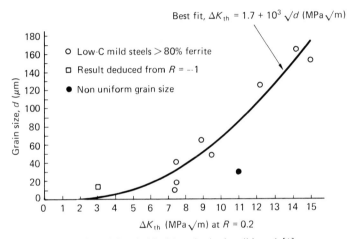

Figure 5.5 Variation of threshold with grain size in mild steels[1]

These data serve to illustrate two points: first, it is clearly necessary to consider different classes of alloys separately; and second, a mechanical parameter such as yield strength may be influenced by different microstructural changes in different systems, so it will generally be more fruitful to consider the effect of microstructure rather than mechanical properties.

Theoretical approaches

A number of theoretical models have been developed which attempt to predict the effect of microstructural parameters; these are discussed in Chapter 2. It can be argued that any attempt to explain near-threshold behaviour must include an element of microstructural control. The principal theoretical models which fall into this category are outlined below.

Yield strength and grain size

A number of theories, approaching the problem by different routes, have predicted similar effects concerning the dependence of grain size and yield strength. In fact, any model which considers crack-tip plasticity, either in terms of plastic zone size, dislocation dynamics or plastic work of fracture, tends to conclude that threshold values *increase* if either grain size or yield strength are increased, other parameters being held constant. The commonest forms for the predictions are:

$$\Delta K_{th} \propto \sigma_y \tag{5.1}$$
$$\Delta K_{th} \propto \sqrt{d} \tag{5.2}$$

A fuller treatment of these theories can be found in Chapter 2; it is, however, interesting to consider how these various approaches come to similar conclusions. Some theorists equate plastic zone size at the threshold to grain size, arguing in terms of a change in the mechanism of crack advance when plasticity becomes localized. More correctly, it is usually the 'knee' or transition point of the growth rate curve at which the change is

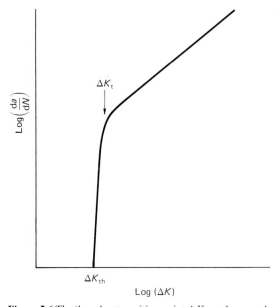

Figure 5.6 The 'knee' or transition point ΔK_t on the growth rate plot

predicted to occur (Figure 5.6). In any case, the threshold will be predicted to depend on the square root of grain size and to be directly proportional to yield strength, through plastic zone size equations of the form:

$$\text{Plastic zone size} \propto \left(\frac{\Delta K}{\sigma_y}\right)^2 \tag{5.3}$$

Other workers assume that crack propagation is controlled by propagation of dislocations along a slip band from the crack tip to the grain boundary.

Again, increasing yield strength tends to make this propagation more difficult and increasing grain size can be beneficial because it increases the amount of energy in the slip band, or the distance over which dislocations have to move. Models of this kind tend to conclude a grain size dependence of the form:

$$\Delta K_{th} = A + B \sqrt{d} \tag{5.4}$$

where A and B are constants. Finally, some workers place emphasis on the non-contiguous advance of the crack front, noting that some parts of the crack front advance faster than others, leaving ligaments of material connecting the crack faces, which have to be fractured at some stage. Figure 2.7 illustrates this concept. In this model, fracture of the ligaments is taken to be the critical step; this is clearly hindered by a higher yield strength in the material, since the ligment fracture is basically low-cycle fatigue. Grain size control can be incorporated by considering that the ligament size is related to the grain size.

At first sight the yield-strength dependence seems to contradict the results shown in Figures 5.3 and 5.4, but it must be remembered that in most cases, yield strength and grain size are linked. For the normal Hall–Petch situation, relevant to single-phase materials and others under grain size control:

$$\sigma_y = \sigma_o + K\sqrt{d} \tag{5.5}$$

it can easily be shown that in order to increase yield strength, grain size must be decreased in such a way that the net effect is a drop in ΔK_{th}. This effect has been demonstrated for a range of materials including mild steels[2] an Fe–C–Ti alloy[3] and Ti–6Al–4V[4].

As will be shown below, the same effect will apply to microstructures in which yield strength is controlled by lath width, packet size, etc., provided the dependence is similar to or stronger than the Hall–Petch dependence. Hence the effect on martensitic/bainitic steels noted in Figure 5.3, for which lath width is expected to be the controlling factor.

In rare cases it is possible to increase yield strength without altering microstructure, for instance by solid solution additions[5], and an increase in threshold results. Difficulties arise in the case of materials which are strengthened within the grains by various kinds of precipitates, especially the aluminium alloys, of which more will be said below.

The use of 'effective grain size'

The above remarks apply strictly only to materials in which plasticity is controlled by grain boundaries, i.e. 'Hall–Petch' materials. However, it might be extended to include any microstructural barriers to plasticity. This concept has been investigated extensively by Yoder and co-workers in many publications (e.g. [6–9], see also Chapter 2). An obvious extension into the small-scale microstructures is to replace grain size with, for example, lath width in martensitic or bainitic microstructures. Yoder *et al.* define a parameter, $\bar{l}$, the effective grain size, defined by the 'mean free path' in the microstructure in question.

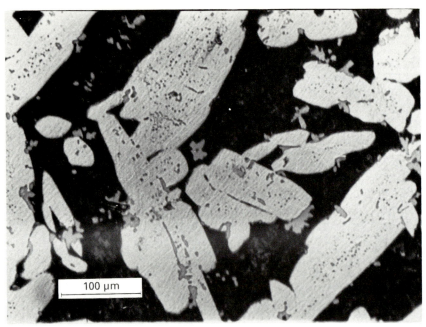

Figure 5.7 A coarse, two-phase material. This quenched Cu–Al alloy has been etched to show Cu-based grains (light phase) and a martensite (dark phase) as well as other precipitates

Problems arise in the case of materials such as pearlite, in which yield strength receives contributions from microstructural elements on more than one scale. It is not clear, in the absence of direct observations, how far plasticity will extend from a crack tip, and which microstructural feature will effectively form the barrier.

Coarse two-phase materials such as ferrite/pearlite and ferrite/martensite present similar problems. Figure 5.7 illustrates one such two-phase material. It is surprising that the Hall–Petch approach can be shown to describe yield strength in these materials up to quite large compositions of the hard phase[10], and this suggests that grain size might be the dominant factor under threshold conditions also. However, for a reasonably large proportion of the hard phase, it might be expected that the spacing of hard phase regions could control plasticity, giving a larger effective grain size. Too great a proportion of hard phase may create a continuous path for the crack through the hard phase with, presumably, a decrease in threshold down to the values which are typical for this phase alone. The morphology of the hard and soft phases has also been shown to play a strong role, probably by altering crack path, as shown by Suzuki and McEvily[11], whose work will be discussed below.

Vaidya[12] has presented results on a ferrite/pearlite steel showing that crack growth rate is strongly influenced by the microstructure at the crack tip; pearlite colonies formed an effective barrier to growth. Figure 5.8 shows the effect of a pearlite colony in causing the crack to branch and deviate. However, it should be noted that Vaidya's crack growth-rate data were gathered from surface observations of cracks and may not represent

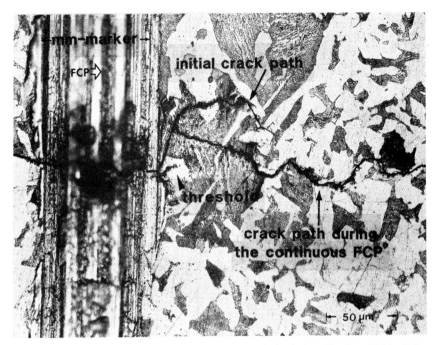

Figure 5.8 Interaction between crack and microstructure in ferrite/pearlite, from Vaidya[12]

the average growth rates of the crack fronts in his material. This highlights a problem inherent in the method of crack measurement by direct observation (see Chapter 4).

One final problem is the distribution of values of grain size, or effective grain size. The grain size of a material is of course an average value, usually derived by a linear intercept method on a two-dimensional section of the microstructure, and often corrected to give a three-dimensional average grain diameter. Strictly speaking, one should also take into account factors such as grain *shape*, which may cause anisotropy in, for example, rolled aluminium alloys. One should also consider the statistical *distribution* of individual grain sizes; for instance, a material which has a very wide distribution about the mean value will contain a relatively large number of grains of extreme sizes (both large and small). Small grains are associated with lower threshold values, therefore the presence of a number of below-average grains might be sufficient to halt the crack, giving a lower threshold than expected. It has certainly been noted that materials with non-uniform grain size distributions show unusual threshold values[1].

Crack deflection

As noted previously, a tortuous crack path, incorporating high angles of local deflection and crack branching, has a lower effective ΔK at the crack tip, due to both mechanics effects and closure effects. Suresh[13–15] has laid the theoretical groundwork for these effects in some detail. Figures 3.6

and 3.7 in Chapter 3 illustrate some actual crack paths which show a high degree of tortuosity. The closure effect is clear in Figure 3.7.

Deflection may be encouraged in a material by various microstructural changes. On the small scale, the introduction of coherent precipitates promotes heterogeneous slip; the formation of well-defined slip bands ahead of the crack encourages faceted growth at relatively high-angle deviations, with the growth angle changing at each grain boundary. This effect accounts for the relatively high thresholds in underaged alloys of aluminium, etc. [16–18].

On a larger scale, the arrangement of phases in a coarse two-phase microstructure such as a dual phase steel has been shown to have strong effects on threshold values. One explanation for this is that the best effects may be achieved if the crack is encouraged to confine itself to the ferrite phase (by constructing a continuous path of ferrite) while at the same time forcing it to deflect through large angles to stay inside this phase; this can be done by having a reasonably large number of islands of hard phase present. Further work on these coarse two-phase microstructures may be rewarding in the future.

Padkin, Brereton and Plumbridge[19] have used finite element analysis to predict the effect of particle *modulus* on crack propagation. Second phase particles and inclusions often have elastic modulii which are very different from the metallic matrix; this analysis suggests that high-modulus particles will tend to retard and deflect cracks, and that low-modulus particles will tend to attract cracks to them, increasing propagation rates. However, the beneficial effect of some low-modulus particles (e.g. graphite in cast iron) may be to trap the crack at the delaminated particle, hindering further growth.

Orientation is also very important, as shown by Mayes and Baker, who have studied the effect of elongated inclusions in free-machining steel[20]. Placing inclusion stringers at right-angles to the crack propagation direction it was possible to achieve thresholds in excess of $14\,\mathrm{MPa}\sqrt{m}$, which is very high for this alloy. Considerable deflection and branching of cracks was noted.

Microstructural control in various alloy systems

This section aims to demonstrate, through examples, possible changes to microstructure which can improve threshold, while noting any effects on other mechanical properties, especially yield strength. Materials are divided up according to alloy system. One general effect, which has been mentioned by a number of workers and which is discussed in Chapter 2, is the variation between one alloy system and another, caused by differences in Young's modulus. This accounts for the generally low ΔK_{th} values in aluminium and titanium alloys, for example, and may set an upper limit to the threshold values which can be achieved in these materials.

Alloy steels

Standard quenched/tempered alloy steels (e.g. Figure 5.9) show a fairly consistent relationship between yield strength and threshold, as shown in

Figure 5.9 Typical quenched and tempered alloy steel microstructure

Figure 5.3, despite the fact that data in this diagram include a wide range of martensitic and bainitic microstructures. It is likely that a detailed study of tempering characteristics would reveal effects similar to 350°C embrittlement which affect crack propagation near the threshold; to my knowledge no detailed study has been conducted, but there is some evidence for a rapid increase in ΔK_{th} for tempering above 400°C[22]. Hippsley[22] showed very little effect of temper-embrittling heat treatments on near-threshold behaviour, though he did not measure to very low growth rates. Figure 5.3 suggests that the benefits to be gained in terms of increased threshold for a constant yield strength are relatively slight, except at the highest yield strengths, where the width of the scatter band is relatively large compared to the ΔK_{th} values involved.

Very little data is available on the threshold behaviour of alloy steels produced by routes other than quenching and tempering, such as the increasingly popular microalloyed steels made by continuous cooling. Farsetti and Blarasin[23] showed that a V–Nb microalloyed steel was as good in the near-threshold region as a typical Cr–Mo steel, for comparable yield strengths up to 800 MPa. It is likely that high thresholds could be achieved, especially from the systems which give rise to lamellar forms of alloy carbide.

Mild steels

The standard normalized mild steels possess very good threshold values, considerably higher than is generally achievable with alloy steels. Mild steels also follow fairly regular rules as regards the relationship between yield strength, grain size and threshold, as illustrated in Figure 5.5, provided the pearlite content is not too high. The work of Mutoh and Radhakrishnan[24] is a typical example of this. One would also expect some morphological effects; for instance Endo and Murakami[25] have shown a strong anisotropy for small cracks growing in a banded-pearlite structure. The effective grain size changed from about 500 μm to 50 μm when crack growth direction changed from parallel to perpendicular, compared to the direction of banding.

Work on completely pearlitic steels has, as yet, been inconclusive[26,27,28]; the data presented in Figure 5.4 suggest that there is no simple relationship between threshold and yield strength, and this is perhaps not surprising since yield strength can be influenced on several microstructural levels. Lloren and Sanchez-Galvez[28] have shown that there is in fact a general trend of decreasing threshold with increasing yield strength; this was shown by testing a very high-strength pearlite, with a yield strength of 2 GPa. Within the normal range of yield strengths, however, there are clearly a number of microstructural variables involved.

Dual-phase steels

Dual-phase steels have attracted much attention in recent years, since it has been shown that they offer the possibility of high thresholds combined with high yield strengths[11,29–32]. However, it should be noted that some dual-phase steels, especially the higher-strength forms, demonstrate poor thresholds; careful microstructural control seems to be necessary. Figure 5.10 compares data for traditional alloy steels (from Figure 5.3) with some results for dual-phase steels.

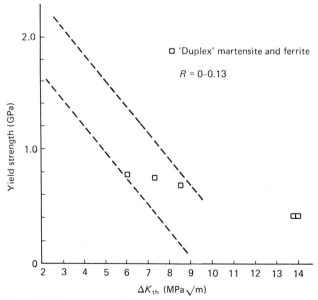

Figure 5.10 Comparison of thresholds in dual phase steels (data points) with typical quenched-and-tempered alloy steels (scatter band from Figure 5.3)

The essential features of a dual-phase steel are ferrite grains and martensite colonies; this microstructure is most simply achieved by cooling rapidly from the ferrite/austenite two-phase region of the phase diagram, but considerable morphological changes are made possible by more complex heat treatments and heat/work cycles.

Possible mechanisms for the advantageous threshold behaviour of these alloys are crack-tip deflection and high effective grain size, as discussed above. It has been suggested[1] that some dual-phase steels are successful because they increase the effective yield strength of the material, while maintaining the same grain size, equal to the ferrite grain size. This is substantiated by the fact that both mild steels and dual-phase steels conform to the same plot of threshold against the parameter, $\sigma_y \sqrt{d}$, shown in Figure 5.11, which is said to be the controlling parameter on a crack-tip

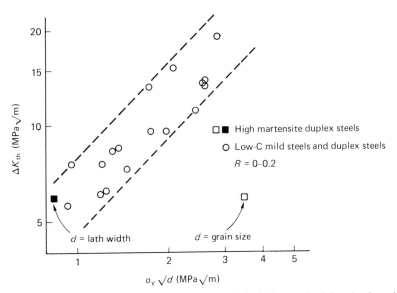

Figure 5.11 The similarity of threshold behaviour in dual-phase and mild steels when plotted against the parameter $\sigma_{y}\sqrt{d}$

plasticity model (see above and Chapter 2). The mechanism for this is that the martensite in the dual-phase steel, occurring as it does on the ferrite grain boundaries, will tend to improve their effectiveness as barriers to slip, while keeping the same grain size. Morphologies consisting of ferrite islands surrounded by martensite tend to give better properties than martensite islands surrounded by ferrite[32]. Too large a proportion of martensite will result in a microstructure which is really a martensite matrix with occasional ferrite grains. In such a case the material should be treated more like a fully martensitic microstructure; Figure 5.11 shows that a high-martensite dual phase steel falls into the same scatter band only if the value of d is replaced by martensite lath spacing (1 μm).

Cast irons

Surprisingly little work has been done on these materials, considering the range of applications for which defect tolerance is required of them, and considering also the range of microstructural variations which are available. Greenan[33] has tested a grey cast iron, obtaining threshold values typical of mild steel; however, work on nodular cast irons such as in Figure 5.12 has demonstrated higher thresholds[34,35], e.g. 11 MPa $\sqrt{m}$ at $R = 0$, with a closure-free $\Delta K_{\mathrm{eff,th}}$ value of 7.8 MPa $\sqrt{m}$[34]. Mechanisms include increased crack deflection (at graphite nodules and porosity) and the closure contribution of graphite debris[35].

Aluminium alloys

Aluminium alloys tend to have much lower values of ΔK_{th} than steels or nickel-base alloys, an effect which may be attributed to a lower value of Young's modulus (see Chapter 2). The extensive use of aluminium alloys in the aerospace and transport industries has generated a lot of data, especially on 7000-series alloys.

Microstructural effects tend to be dominated by precipitate size and morphology; indeed there are almost no results available on the effect of grain size. The existence of a grain size effect similar to that described above is, however, hinted at by the work on an extremely fine-grained material reported by Minakawa, Levan and McEvily[36], which had a very low threshold.

An effect which has been noted by a number of workers[16–18,38] is the relative success of underaged alloys, possessing coherant precipitates which encourage heterogeneous slip and therefore aid crack deflection (see above). This means that the threshold tends to decrease continuously as ageing proceeds, despite the rise and fall of yield strength, offering the possibility of optimizing ΔK_{th} and yield strength by careful tempering.

Vasudevan, Bretz and Miller[39] note the superior properties of the Al–Li alloy 2020 over a more traditional aerospace alloy, 7075. This paper is valuable since Al–Li alloys have gained much interest in aerospace applications; the paper is also useful from a mechanistic point of view because it considers a range of possible mechanisms, including oxide closure, deflection and microstructure. Defelection is advanced as the key effect in promoting high thresholds in this alloy, though in fact two other effects seem to be present. First, oxide thickness seems to be great enough to induce an oxide-closure effect (0.1 µm of oxide was measured on each face near threshold, where the CTOD was predicted to be 0.2 µm); second it was noted that the material had a large recrystallized grain size which, since the yield strengths of the two alloys are similar, might be expected to increase ΔK_{th}.

In general, however, it seems that oxide-induced closure does not play a role in these alloys; the principal closure mechanism being roughness, caused by deflection.

Edwards and Martin[40] have demonstrated considerable increases in ΔK_{th} combined with modest increases in yield strength, for an Al–Mg–Si alloy, by addition of dispersoids of a manganese-containing intermetallic.

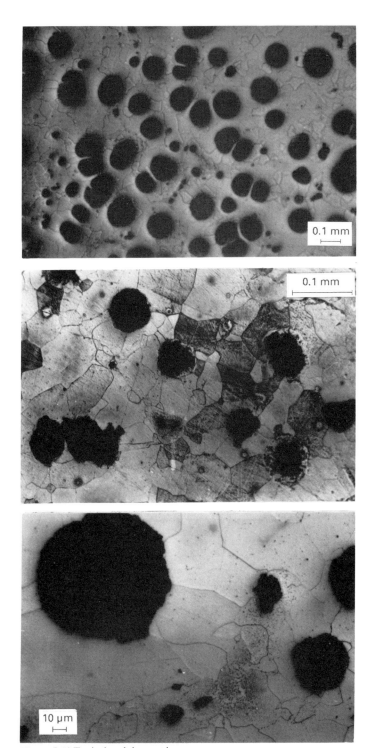

Figure 5.12 Typical nodular cast iron

This alloy is unusual in that it shows significant amounts of static, intergranular fracture, even at the threshold. Additions of dispersoids promoted transgranular failure in the near-threshold region.

An additional complication for most commercial aluminium alloys is the complex nature of the elongated grain structure, arising from cold-rolling, etc., which encourages anisotropy.

Nickel-base alloys

Work on the nickel alloys has concentrated almost exclusively on the complex, commercial alloys intended for elevated-temperature use in jet engines, turbines, heat exchangers, etc. Interpretation of results is therefore difficult owing to the variety of microstructural features present and the lack of comparable data on simpler nickel alloys.

A very good review has been written by King[41], covering in detail the known effects of microstructure at room temperature and elevated temperatures. Thresholds are generally found to increase with increasing grain size[42,43], as shown in Figure 5.13, though the point at which the plastic zone size becomes equal to the grain size occurs at a much higher stress intensity than the threshold itself. Thresholds also increase with a decrease in precipitate size, which seems, as with the aluminium alloys, to be related to a change in slip character. As Figure 5.14 shows, deformation may be confined to intense slip bands, or may be relatively homogeneous.

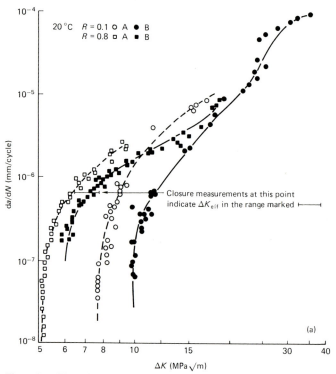

Figure 5.13 The effect of grain size and R ratio in nickel-base alloys, from King[41]

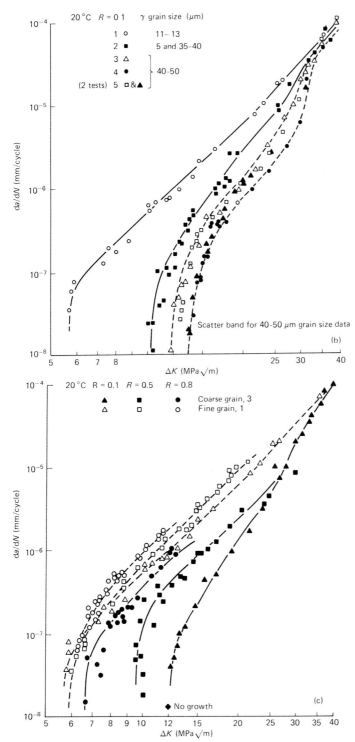

Figure 5.13 (contd)

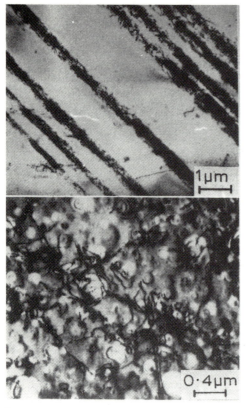

Figure 5.14 Homogeneous and heterogeneous slip character in nickel-base alloys, from King[41]

Heterogeneous deformation tends to increase surface roughness and therefore enhance roughness-induced closure.

Behaviour at high temperatures is complex, but a general trend towards decreasing thresholds (by up to 40%) seems to be linked to this change in slip character also, though up to 25% of the decrease can be attributed to the gradual decrease in both yield strength and elastic modulus as temperature is increased. Figure 5.15, taken from King's review[41], illustrates very well the changes in fracture surface appearance and surface roughness which occur as grain size and temperature are varied.

Titanium alloys

A very wide range of microstructural types are possible in titanium alloys, ranging from almost single-phase α grains to various quenched/tempered martensitic α/β structures. Microstructural effects in these alloys have been studied extensively by Yoder *et al.* [6–9] and others[44–46]. Ti–6Al–4V is by far the most extensively studied, and has the distinction of being one of the few commercial alloys to be used in a Widmanstatten form; even fully quenched material may have Widmanstatten structures in the centre of

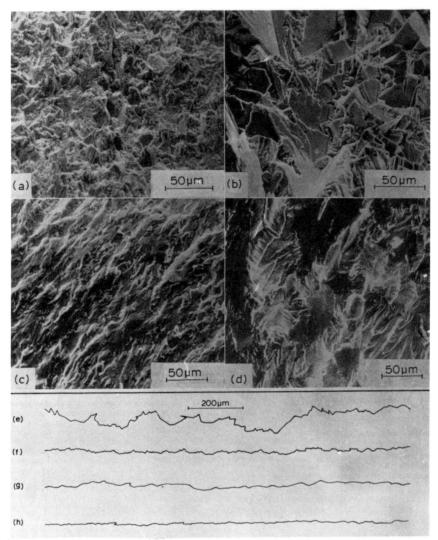

Figure 5.15 Changes in fracture surface appearance and surface roughness resulting from changes to grain size and temperature in nickel-base alloys, from King[41]: (a, f) 5–12 μm grain size, 20°C; (b, e) 50 μm grain size, 20°C; (c, h) 5–12 μm grain size, 600°C; (d, g) 50 μm grain size, 600°C

thick sections, and β-transformed conditions can be almost 100% Widmanstatten α.

These Widmanstatten forms tend to have the highest threshold values[44], giving good combinations of ΔK_{th} and yield strength. Typical grain size effects can be observed in pure α and α/β microstructures[45].

On the whole it can be said that the titanium alloys overcome the problem of an inherently low Young's modulus far better than the aluminium alloys, due to their versatility under thermomechanical treatment. This makes them extremely attractive in applications requiring optimal values of specific strength, specific stiffness and fatigue threshold.

Stainless steels

Austenitic stainless steels have moderate threshold values, typically 6-7 MPa $\sqrt{m}$ at $R = 0$. This can be varied by changes to grain size[24,47], but the majority of commercial alloys possess very similar thresholds. There are, however, very significant changes associated with environmental conditions such as saline solution or vacuum (see Chapter 6).

The ferritic and martensitic stainless steels and iron-based superalloys demonstrate large variations in threshold typical of those discussed above for alloy steels and titanium alloys. For instance, martensitic alloys tend to show an inverse relationship between threshold and yield strength[48–50].

Non-metallic materials

Very little work has been done on the measurement of threshold values in non-metallic materials. It may be argued that threshold fatigue crack propagation behaviour is relatively unimportant in *ceramics*, though slow fatigue crack growth has been measured on, for example, polycrystalline alumina[51] tested under cyclic compression.

Concerning *polymers*, near-threshold behaviour may be life-controlling in many applications, especially in the case of brittle polymers such as PMMA (see Chapter 10); a number of workers have recognized the need to measure da/dN values, but near-threshold results are rare; data by Usami on epoxy resin is one example[52].

Polymer *composites* do not tend to fail by single-crack propagation, but by gradual accumulation of small-scale damage such as delamination cracks, fractured fibres, etc. There is some value, however, in developing LEFM approaches to cover this situation. At present I know of no work on the measurement of thresholds in such composites.

Microstructural effects in short cracks

The unique behaviour of short fatigue cracks is discussed at length in Chapter 8. It is important to consider these two chapters in conjunction, because it is found that microstructure has profound effects on short-crack behaviour which are often quite different from the effects described above, which are confined to conventional long cracks.

Concluding remarks

Microstructural control is a powerful tool in the near-threshold regime, and there is still much potential for alloy design in this area, as exemplified by the dual-phase steels. In particular, alloys can be envisaged which have improved long-crack thresholds without sacrificing properties such as yield strength, creep strength and short-crack fatigue strength.

At a simple level, the general effects of microstructural variables, and of mechanical properties such as yield strength and modulus, can be understood in terms of known mechanisms of crack propagation. However, the behaviour of the more fatigue-resistant microstructures is still difficult to predict from existing theory.

References

1. Taylor, D. (1985) In *Fatigue 84*, EMAS, Warley, UK, p. 327
2. Masounave, J. and Bailon, J. P. (1976) *Scripta Metallurgica*, **10**, 165
3. Benson, J. P. (1979) *Metal Science*, **13**, 535
4. Brown, C. W. and Taylor, D. (1984) In *Fatigue Crack Growth Threshold Concepts*, TMS-AIME, USA, p. 433
5. Robinson, J. L. and Beevers, C. J. (1973) *Metal Science*, **7**, 153
6. Yoder, G. R., Cooley, L. A. and Crooker, T. W. (1980) In *Titanium '80, Proceedings of the Fourth International Conference on Titanium*, p. 1865
7. Yoder, G. R., Cooley, L. A. and Crooker, T. W. (1985) In *Fatigue 84*, EMAS, Warley, UK, p. 351
8. Yoder, G. R., Cooley, L. A. and Crooker, T. W. (1977) *Metallurgical Transactions*, **8A**, 1737
9. Yoder, G. R., Cooley, L. A. and Crooker, T. W. (1979) *Engineering Fracture Mechanics*, **11**, 805
10. Baker, T. N. (1983) *Yield, Flow and Fracture of Polycrystals*, Applied Science, London
11. Suzuki, H. and McEvily, A. J. (1979) *Metallurgical Transactions*, **10A**, 475
12. Vaidya, W. V. (1986) *Fatigue of Engineering Materials and Structures*, **9**, 305
13. Suresh, S. (1983) *Engineering Fracture Mechanics*, **18**, 577
14. Suresh, S. (1984) Report No. MRL E-153, Brown University, USA
15. Suresh, S. (1983) *Metallurgical Transactions*, **14A**, 2375
16. Zaiken, E. and Ritchie, R. O. Report No. UCB/RP/84/A1022, Department of Materials Science and Mineral Engineering, University of California, Berkeley
17. Petit, J. (1984) In *Fatigue Crack Growth Threshold Concepts*, TMS-AIME, USA, p. 3
18. Renaud, P., Violan, P., Petit, J. and Ferton, D. (1982) *Scripta Metallurgica*, **16**, 1311
19. Padkin, A. J., Brereton, M. F. and Plumbridge, W. J. (1987) *Materials Science and Technology*, **3**, 217
20. Mayes, I. C. and Baker, T. J. (1986) *Materials Science and Technology*, **2**, 133
21. Du Bai-Ping, Li Nian and Zhou, Hui-Jiu (1987) *International Journal of Fatigue*, **9**, 43
22. Hippsley, C. A. (1987) *Materials Science and Technology*, **3**, 912
23. Farsetti, P. and Blarasin, A. (1988) *International Journal of Fatigue*, **10**, 153
24. Mutoh, Y. and Radhakrishnan, V. M. (1986) *Journal of Engineering Materials and Technology* (ASME), **108**, 174
25. Endo, M. and Murakami, Y. (1987) *Transactions of the American Society of Mechanical Engineering, Journal of Engineering Materials and Technology*, **109**, 124
26. Gray, G. T., Thompson, A. W., Williams, J. C. and Stone, D. H. (1982) In *Fatigue Thresholds*, EMAS, Warley, UK, p. 345
27. Beevers, C. J. (1982) In *Fatigue Thresholds*, EMAS, Warley, UK, p. 257
28. Lloren, J. and Sanchez-Galvez, V. (1987) *Engineering Fracture Mechanics*, **26**, 869
29. Bulloch, J. H. and Kennedy, R. (1985) *Res Mechanica*, **15**, 259
30. Bulloch, J. H. (1987) *Res Mechanica*, **22**, 325
31. Shang, J. K., Tzou, J. L. and Ritchie, R. O. (1987) *Metallurgical Transactions*, **18A**, 1613
32. Ramage, R. M., Jata, K. V., Shiflet, G. J. and Starke, E. A. (1987) *Metallurgical Transactions*, **18A**, 1291
33. Greenan, A. F. (1982) In *Fatigue Thresholds*, EMAS, Warley, UK, p. 363
34. Griswold, F. D. and Stephens, R. I. (1987) *International Journal of Fatigue*, **9**, 3
35. Ogawa, T. and Kobayashi, H. (1987) *Fatigue and Fracture of Engineering Materials and Structures*, **10**, 273
36. Minakawa, K., Levan, G. and McEvily, A. J. (1986) *Metallurgical Transactions*, **17A**, 1787
37. Edwards, L., Busby, A. K. and Martin, J. W. (1986) *Materials Science and Technology*, **2**, 823
38. Ritchie, R. O., Yu, W., Blom, A. F. and Holm, D. K. (1987) *Fatigue and Fracture of Engineering Materials and Structures*, **10**, 343

39. Vasudevan, A.K., Bretz, P. E., Miller, A. C. and Suresh, S. (1984) *Materials Science and Engineering,* **64**, 113
40. Edwards, L. and Martin, J. W. (1983) *Metal Science,* **17**, 511
41. King, J. E. (1987) *Materials Science and Technology,* **3**, 750
42. Krueger, D.D., Antolovich, S.D. and Van Stone, R. H. (1987) *Metallurgical Transactions,* **18A**, 1437
43. King, J. E. (1989) *Metal Science,* **16**, 345
44. Chesnutt, J. C. and Wert, J. A. (1984) In *Fatigue Crack Growth Threshold Concepts,* TMS-AIME, USA, p. 83
45. Brown, C. W. and Taylor, D. (1984) In *Fatigue Crack Growth Threshold Concepts,* TMS-AIME, USA, p. 83
46. Hicks, M. A., Jeal, R. H. and Beevers, C. J. (1983) *Fatigue of Engineering Materials and Structures,* **6**, 951
47. Priddle, E. K. (1978) *Scripta Metallurgica,* **12**, 49
48. Lindley, T. C. and Richards, C. E. (1982) In *Fatigue Thresholds,* EMAS, Warley, UK, p. 967
49. Moren, K. E. (1975) *Scandanavian Journal of Metallurgy,* **4**, 255
50. Lou, B. and Averbach, B. L. (1983) *Metallurgical Transactions,* **14A**, 1899
51. Ewart, L. and Suresh, S. (1986) *Journal of Materials Science, Letters,* **5**, 774
52. Usami, S. (1982) In *Fatigue Thresholds,* EMAS, Warley, UK, p. 205

6 The effects of environment and temperature

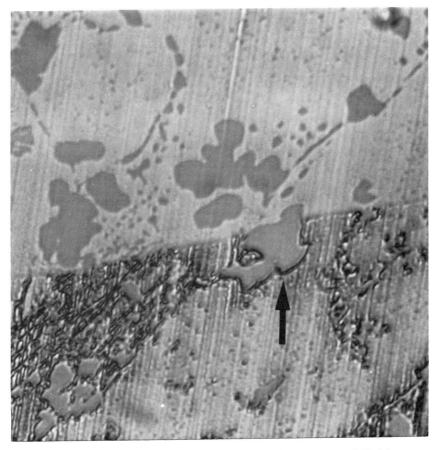

Preferential corrosion on the interfaces of second-phase particles. The upper half of the surface was protected from corrosion, the lower half exposed to sodium chloride solution

Introduction

Crack propagation rates in general, and near-threshold behaviour in particular, can be strongly influenced by the surrounding environment and the prevailing temperature.

It will be shown below that environmental effects are very alloy-specific, so that conclusions drawn from one alloy class may not hold for other alloys. The discussion of temperature effects will be confined to moderate temperatures for which a modified ΔK-based approach may be valid. High temperatures at which creep crack growth dominates are beyond the scope of this book. The effect of temperature reduction, below ambient temperature, is also discussed.

Environment

Mechanisms

A complete description of the effects of corrosive environments on fatigue behaviour would be extremely complex, owing to the large number of mechanisms involved and their dependence on alloy type, thermomechanical treatment and loading conditions. However, under near-threshold conditions, three mechanisms seem to dominate behaviour:

1. Crack closure: the principal effect here is the increase in closure resulting from creation of voluminous deposits on the crack faces. This

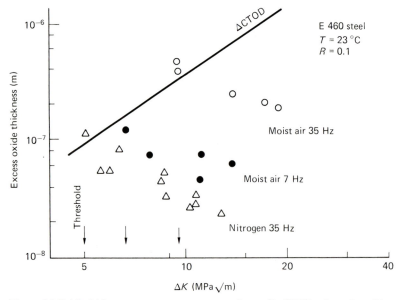

Figure 6.1 Oxide thickness measurements compared to cyclic CTOD values, from Bignonnet *et al.* [1]

may be a purely chemical effect, but more commonly it is amplified by fretting contact between the surfaces which tends to increase the thickness of the deposit by continually revealing fresh metal surface. The most common species of deposit is oxide; oxide-induced closure is a very important mechanism, especially in steels (see Chapter 3). Figure 6.1 shows measurements of oxide thickness compared to cyclic crack opening displacement (CTOD), illustrating the effect of moist air as an oxidizing environment[1]. Note also the effect of frequency in this plot, which is discussed below.

In theory, corrosive effects might be expected to reduce closure by dissolving away asperities on the crack faces, if the electrochemical conditions were appropriate for metal polishing, but this effect has not been reported as far as the author is aware.

Another, less important, environmental effect is the increase in closure brought about by viscous fluids (see Chapter 2 on closure), which would be effective whether or not the fluid in question was corrosive.

2. Standard corrosion-fatigue mechanisms, such as hydrogen embrittlement and crack-tip dissolution, which are known to operate in many alloy systems over a large range of stress intensity values: these may also be expected to have an influence at the threshold, unless they are related to a stress-corrosion-cracking (SCC) threshold, K_{Iscc}, which is higher than the value of K_{max} at the threshold.

It is useful to define three types of corrosion fatigue crack growth behaviour[2], as illustrated in Figure 6.2:

(a) 'True' corrosion fatigue in which SCC does not occur
(b) Stress corrosion cracking occurring under cyclic loading, with no other corrosion fatigue mechanism
(c) Both of the above effects.

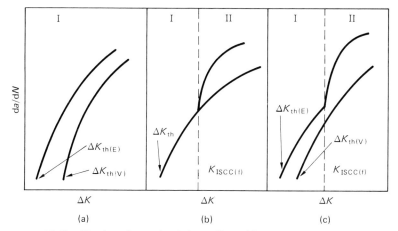

Figure 6.2 Classification of corrosion fatigue effects; (a) 'true' corrosion fatigue, no SCC; (b) SCC only; (c) both 'true' corrosion fatigue and SCC

Corrosion fatigue mechanisms under (a) and (b) above will tend to lower the threshold, and therefore are usually working in opposition to the closure effects.

3. The effect of oxidation on slip character: it has been argued recently[3] that oxidation decreases the reversibility of slip and so tends to accelerate damage near the crack tip. This mechanism will work in opposition to the closure effects which are known to occur in oxidizing environments in some alloys. In principle, similar effects may result from other gaseous corrosive environments, such as carburizing or nitriding.

Crack deflection

It was shown in Chapter 3 that deflection and branching in the crack path had a strong effect on near-threshold behaviour through the increase in closure levels for the relatively rough crack faces. Changes of environment, as well as of alloy type and heat treatment, may have a strong effect on the crack path, and therefore on the degree of roughness-induced closure[4]. Possible mechanisms for this include changes in slip character and the presence of intergranular embrittlement. In addition to closure effects, crack deflection and branching tends to reduce crack-tip ΔK levels, as shown by Suresh[5].

Frequency effects

It is well known that corrosion fatigue behaviour is strongly affected by applied frequency, especially for aqueous corrodants for which the chemical reaction rates are relatively slow compared to gaseous corrodants. The common observation is that reduction of the frequency in the range 100–1 Hz increases fatigue behaviour, whether measured as crack propagation rate[6,7] or as number of cycles to failure. Below 1 Hz the situation may improve again[7], suggesting that in many aqueous systems there is an optimum strain rate for environmental interaction (Figure 6.3). Waveform can also have a strong effect, presumably through changes to the applied strain rate; for instance, square wave cycling has been shown to be less deleterious than sine wave cycling[6].

Unfortunately, very few studies of low-frequency effects have included the threshold region, owing to the prohibitively long testing times involved. Three recent studies on the effect of NaCl solution on near-threshold behaviour[9–11], though not totally in agreement, have suggested the worrying possibility that the thresholds may be significantly lower than those measured at normal testing frequencies. The results of these studies will be examined in more detail below. This raises an important point on the subject of threshold measurement; since most tests are necessarily conducted at high frequencies, above 10 Hz, the results many be non-conservative if frequency effects tend to lower threshold values.

A further worrying effect of frequency has recently emerged as a result of studies on oxide-induced closure[12–14]. It has been shown that oxide build-up on crack faces is greatly reduced at low frequencies. In some

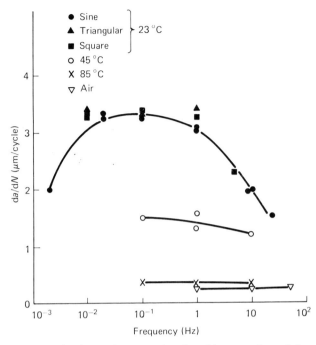

Figure 6.3 Crack growth rate as a function of frequency in steel, from Smith and Stewart[7]

cases, materials which showed strong effects of oxide closure at 100 Hz showed no effect at all at frequencies below 10 Hz. This is illustrated above in Figure 6.1, comparing oxide build-up at 7 Hz and 35 Hz in a moist air; assuming that the threshold occurs when the oxide thickness becomes equal to the predicted CTOD, the diagram predicts a dramatic effect of frequency on threshold. Both steels and aluminium alloys demonstrate this effect. Therefore, material used in service at low-frequency loading may exhibit a much lower threshold than may be deduced from higher-frequency laboratory testing. The mechanism for this latter frequency effect is not clear at present, but it may be due to the fact that, at high frequencies, local heating may raise the temperature near the crack faces, enhancing oxide formation. Useful parallels may be drawn with results from tribological studies of oxide build-up on bearing surfaces[15–17]. Effects of load level and sliding contact velocity noted in these studies may be related to effects of stress intensity and frequency, respectively, in fatigue.

Pseudo-thresholds at higher growth rates

Figures 6.4 and 6.5 illustrate another problem associated with the measurement of thresholds under corrosive conditions. Figure 6.4 presents some very comprehensive results from Bamford[18] which show that corrosive effects at high ΔK levels, caused by NaCl and high relative

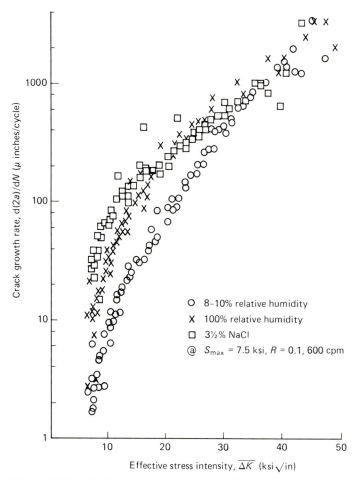

Figure 6.4 Corrosion fatigue data due to Bamford[18]

humidity, seem to disappear at the threshold in this material, becoming coincident with the low-humity results. It should be noted, however, that the da/dN values here are all very high; the 'knee' of the corrosion fatigue data is around 100 microinches per cycle, or 2.5×10^{-3} mm/cycle, whereas near-threshold effects normally set in around 10^{-6} mm/cycle.

The data in Figure 6.5, due to Ward-Close and Beevers[19] shows what may be happening in Bamford's case. Although the number of data points in this case is small, it is clear from the data at the lowest growth rates that the rapid decrease in da/dN which sets in at around 3×10^{-8} m/cycle (3 x 10^{-5} mm/cycle) is not in fact the true threshold, but a temporary 'step' which levels out again at lower growth rates. Presumably, the threshold for this material is lower, around 2–3 MPa $\sqrt{m}$, but this was not measured here.

This 'pseudo-threshold' behaviour occurs in a variety of systems; in some cases it may be related to the stress-corrosion cracking threshold, K_{Iscc}, but may occur in systems not susceptible to SCC. It is important in all cases to

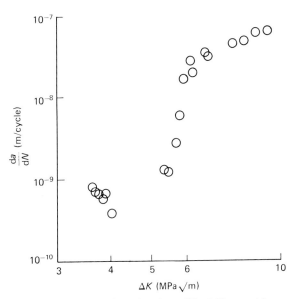

Figure 6.5 Corrosion fatigue data due to Ward-Close and Beevers[19]

establish thresholds at growth rates below 10^{-7} mm/cycle to avoid being misled as to both the absolute value of ΔK_{th} and the effect of a given corrosive environment.

Environmental effects considered by alloy type

In considering the detailed effects of various environments, and their relationships to the prevailing mechanisms mentioned above, it is convenient to divide up materials by alloy class. The majority of the data discussed below, apart from the most recent results, may be found in collected form in[20].

Steels

A useful general rule, which applies to almost all the results generated on steels, is that normal laboratory air produces a higher threshold value than any other environment, including the more inert environments. This is unfortunate in the sense that laboratory-air data is non-conservative with respect to changes in environment.

Relatively inert environments such as a vacuum[21,22] and dry inert gases[23,24–29] cause threshold reductions by removing the effect of oxide-induced closure. More aggressive aqueous corrodants, such as NaCl solution, might be expected to show enhanced closure effects for the same reason, but in fact it seems to be the classic mechanisms of hydrogen embrittlement and crack-tip dissolution which dominate, tending to accelerate crack growth rates and depress thresholds. Figure 6.6 typifies the extensive results of Ritchie, Suresh and co-workers[22,24–27], who

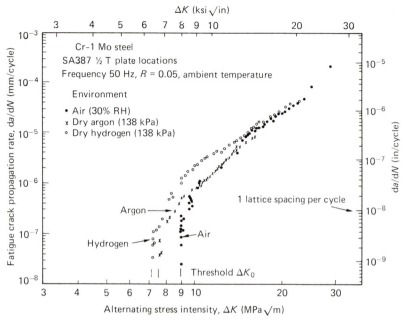

Figure 6.6 Typical effect of gaseous environments in steels, Ritchie, Suresh and Moss[23]

demonstrated low threshold values for many gaseous environments which had been carefully dried to prevent oxidation. Figure 6.7 illustrates the behaviour of a mild steel in NaCl solution[51]. A typical mild steel would have a threshold of $7\,\mathrm{MPa}\sqrt{m}$ in air at a low R ratio, but this value may drop as low as $2\,\mathrm{MPa}\sqrt{m}$ as a result of environmental effects, though values in the range $4-5\,\mathrm{MPa}\sqrt{m}$ are more usual. At high R ratios, closure effects disappear, so inert environments tend to lose their effectiveness.

Some exceptions to this general rule have been recorded. For instance, some workers have shown beneficial effects of aqueous solutions as a result of oxide or other deposits inducing closure[10,31]. Also, in some cases, a vacuum has been reported to increase the threshold[1,32,33]; a mechanism for this may be crack-tip rewelding, as noted by Hippsley[33]. In all cases involving inert gaseous environments, the presence of trace elements, especially traces of water vapour, can be shown to have profound effects through catalytic control of oxidation and other gas/metal reactions. Even very active gases such as hydrogen and oxygen can be rendered relatively ineffectual if all traces of water are excluded by careful drying. This tends to produce considerable differences between the results of different researchers; accurate gas analysis equipment is essential for this type of experimental work.

Smith[34] has conducted a useful study which demonstrates the effectiveness of oxide-induced closure in an alloy steel. He was able to obtain dramatic changes in crack growth-rate and closure levels in individual specimens by changing from dry to moist air atmospheres during the test. An interesting note is that, when the atmosphere was changed

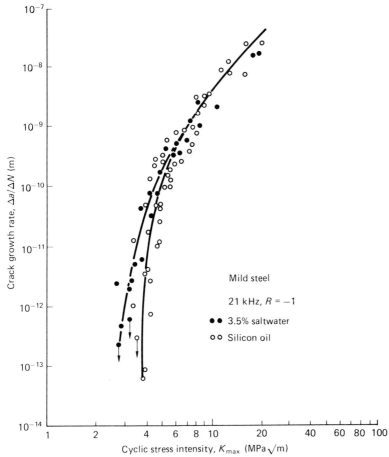

Figure 6.7 Typical effect of sodium chloride solution on mild steel, from Stanzl and Tschegg[51]

from dry to moist, it took more than 0.5 mm of crack growth to establish the new closure levels and oxide layers.

Aluminium alloys

The typical effect of inert environments is well illustrated by results presented in Figure 6.8[13]. For both underaged (UA) and overaged (OA) versions of the 7075 alloy, the effect of vacuum is to raise threshold and also to keep it constant to variations in R ratio[35–38]. For the overaged alloy the vacuum-derived threshold has the same value as the low-R threshold in air, whereas for the underaged alloy the threshold in vacuum is considerably higher. The vacuum/air comparison here implies that oxide closure effects must be unimportant in these alloys, though direct measurements of oxide layers present conflicting evidence[39–41].

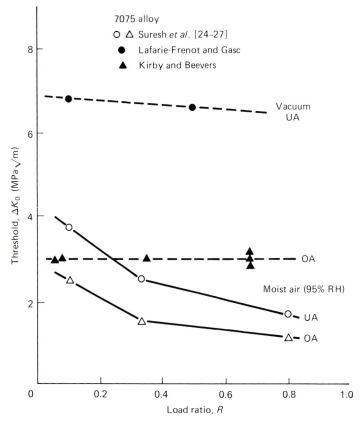

Figure 6.8 Typical results on normal and inert environments for aluminium alloys[13]

Relatively wet inert gas environments can show very low thresholds[4,42], though not always[43]. This suggests that a classic corrosion-fatigue mechanism such as hydrogen embrittlement[42] may be dominant. Aqueous corrodants may lower the threshold[11], especially when tested at low frequencies, as shown in Figure 6.9, though the effects are less dramatic than in ferrous alloys, as might be expected from the increased resistance to normal corrosion processes.

A difficulty with the study of data on aluminium alloys is that the great majority of results have been generated on a few systems, notably the 7000 series alloys. This arises naturally out of their applications in aerospace, etc., but for the researcher this tends to produce a biassed analysis, the more so since these commercial alloys are relatively complex in composition and microstructure.

Other non-ferrous alloys

Data on environmental effects at room temperature are much more scarce for other alloy systems than they are for the ferrous and aluminium alloys. For the copper, nickel and titanium alloys the general rule seems to be, as

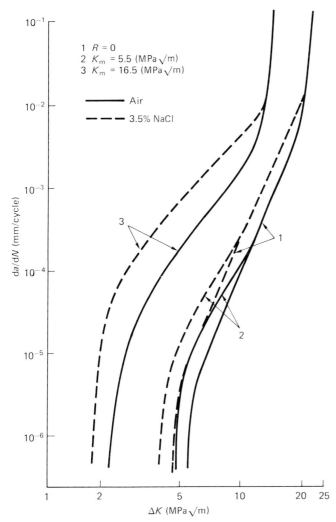

Figure 6.9 Results due to Radon[11], testing aluminium alloys in aqueous environments at relatively low frequencies

for aluminium alloys, that inert environments increase threshold[44–49] and corrosive environments (aqueous and gaseous) decrease threshold[50,51]. Figure 6.10 illustrates the deleterious effect of hydrogen in a titanium alloy at various temperatures[52]; Figure 6.11 shows the effect of NaCl solution in reducing the threshold of copper[51]. This latter effect is surprising for a relatively strong oxide-former such as copper, for which the dominant effect might be expected to be a reduction in threshold after oxide-induced closure. Again, classic mechanisms such as hydrogen embrittlement or dissolution seem to be operating; however, more data would be needed on these materials before definate conclusions can be drawn.

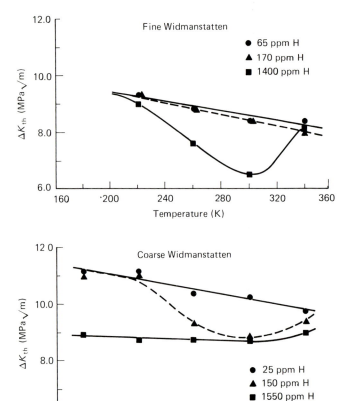

Figure 6.10 Effect of hydrogen on a titanium alloy at various temperatures [52]

The influence of crack closure

A general point which emerges from the above results is that mechanisms based on crack closure have relatively little influence on environmental effects outside the ferrous system, despite the relatively widespread interest which has been generated in the study of this mechanism and the concomitant decrease in work on the traditional corrosion-fatigue mechanisms.

Temperature

Relatively little work has been done on the changes to ΔK_{th} caused by elevated or reduced temperatures, except in the case of certain alloy systems which are designed for high-temperature use, notably nickel-base superalloys and certain alloy steels.

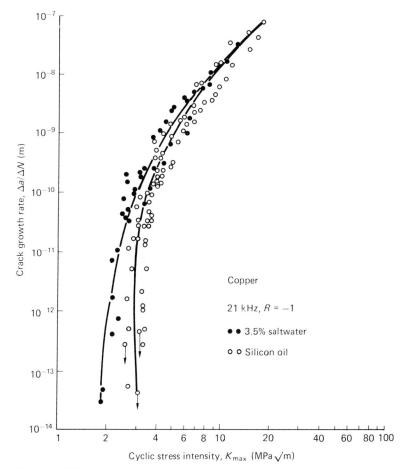

Figure 6.11 Effect of sodium chloride solution on copper[51]

At sufficiently high temperatures, creep crack growth will dominate, and it will generally not be possible to use ΔK as a characterizing parameter. The description of creep crack growth and of creep-fatigue interactions is beyond the scope of this book, therefore this section will be confined to moderate temperature elevations which may cause changes to the value of ΔK_{th} without violating LEFM conditions. Possible mechanisms which can be envisaged to cause temperature effects are:

1. Oxide-induced closure: the enhancement of oxidation rates might be expected to cause more closure, and therefore lower thresholds, at elevated temperatures.
2. Corrosion processes which are rate-controlled might be expected to increase with temperature, lowering ΔK_{th} if, for example, crack-tip dissolution operates, but increasing threshold if a voluminous deposit causes closure, as above.

3. Embrittlement may be affected by temperature changes, either by altering the kinetics of an embrittlement process or by moving through a ductile/brittle transition in the material.
4. Changing temperature will change yield strength: even in the absence of microstructural changes the yield strength will tend to decrease with increasing temperature in most materials. The effect of yield strength on ΔK_{th} is complex, but most theoretical models suggest that ΔK_{th} is approximately proportional to yield strength in the absence of microstructural changes (see Chapter 2). If this is the case, then increasing temperature should increase threshold.
5. Microstructural changes, such as precipitate coarsening or phase changes, will affect properties in a complex manner.

Data on temperature effects

As with the preceding section on environmental effects, it is convenient to consider temperature effects in terms of alloy type.

Ferritic steels

A relatively large amount of evidence exists to show that increasing temperature tends to lower threshold, and decreasing temperature below ambient tends to increase threshold[53–64]. This applies for both mild steels and ferritic alloy steels. A typical example of the behaviour of a low-alloy steel with relatively good high-temperature properties is given by the work of Puskar and Varkoly[63]. Here a gradual decrease in ΔK_{th} with increasing temperature was noted, the value of ΔK_{th} at 500°C being approximately 50% of its room-temperature value. Figure 6.12 shows typical low-temperature results for a mild steel[63].

This suggests that the dominant effect is the change in yield strength and dislocation mobility resulting from thermal activation, rather than mechanisms such as oxide-induced closure or coarsening of carbide phases. Other mechanisms which might lower threshold in steels under moderate heating, in the range 100–300°C, are dynamic strain ageing and reduction of water-vapour content which enhances oxide formation.

One piece of work contrasts with the general findings[58]; here a CrMoV steel tested at 550°C showed high thresholds which tended to increase with decreasing frequency. However, the near-threshold region was very poorly defined in this work.

Non-ferrous materials, superalloys

Nickel-base and iron-base superalloys have received considerable attention as regards their elevated temperature thresholds[44,65–68]. These materials, as would be expected, perform well up to relatively high temperatures. Above 700°C, results are complex, with strongly different effects shown by different alloys, and even by different thermomechanical conditions of the same alloy. Large increases and decreases in threshold have been observed when the temperature is raised above 800°C,

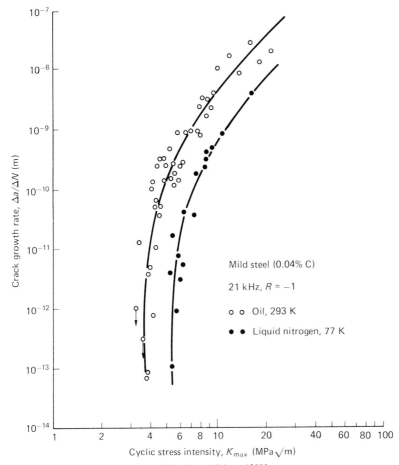

Figure 6.12 Typical temperature effects in a mild steel[53]

presumably due to the onset of creep effects (crack growth or crack-tip blunting). Interestingly, the lowest thresholds have been measured at elevated temperature in vacuum.

For temperatures up to 600°C, which constitutes the working temperature range for most jet engine alloys, the general effect is a gradual decrease in threshold by 25–40% [68]; up to 25% of this may be attributed to decreases in yield strength and modulus, but, as King[68] has noted, thresholds may also decreased through changes to slip character at elevated temperatures, which tend to reduce crack closure. Relatively high thresholds sometimes occur at certain elevated temperatures, despite an increase in crack growth rates at higher stress intensities[69]; this effect may be linked to oxide-induced closure.

Undoubtedly, some of the complexity of the results presented must be caused by the practical difficulty of measuring slow crack growth rates at these high temperatures.

Considering low-temperature behaviour, Liaw and Logsden[70] have measured thresholds in Inconel 706 at the lowest temperature normally attainable, i.e. 4.2 K, in liquid helium. They reported a slightly higher threshold than that measured in ambient air, at $R = 0.1$, with an absence of R ratio effect at the low-temperature leading to a much bigger difference when the two temperatures were compared at $R = 0.8$. They noted that the ambient-air specimens showed more closure, linked to more surface roughness than the low-temperature specimens, which would tend to have the opposite effect on threshold. Dislocation mobility was suggested as a possible reason for the favourable results at the low-temperature; in fact the reason may have been more connected with the fact that the low-temperature environment was an inert gas.

Concerning austenitic stainless steels, elevated temperature has been reported both to decrease and increase threshold[30,32,48]; the results of Usami, Fukuda and Shida show particularly strong increases for temperatures up to 550°C in SUS304. One report on a copper alloy[53] showed a slight increase in threshold when the material was cooled to liquid nitrogen temperature.

Since many of the mechanisms listed above are expected to be time-dependent, including microstructural changes and creep, one must expect significant effects of frequency and testing procedure, since some testing procedures take much longer periods to reach threshold than others. The measurement of near-threshold crack propagation rates at elevated temperatures is difficult, and few experimental facilities are available for this type of work (see Chapter 4 on methods of measurement).

Conclusions

Environmental and temperature effects are still poorly understood and remain a major stumbling block to the use of threshold concepts in certain applications. At present the complexity of the problem, the number of competing mechanisms involved and the liklihood of frequency effects and other testing problems, suggest that the only realistic approach is to test the material in conditions which are as close as possible to service conditions, if environmental effects are suspected.

References

1. Bignonnet, A., Loisin, D., Namdir-Irani, R., Bouchet, B., Kwon, J. H. and Petit, J. (1983) In *Fatigue Crack Growth Threshold Concepts*, TMS-AIME, USA, p. 99
2. Gerberich, W. W. and Yu, W. (1982) In *Fracture Problems and Solutions in the Energy Industry*, Pergamon, Oxford, p. 39
3. Suresh, S. and Ritchie, R. O. (1983) in *Fatigue Crack Growth Threshold Concepts*, TMS-AIME, USA, p. 227
4. Petit, J. (1983) In *Fatigue Crack Growth Threshold Concepts*, TMS-AIME, USA, p.3
5. Suresh, S. (1983) *Metallurgical Transactions*, **14A**, 2375
6. Taylor, D. and Knott, J. F. (1982) *Metals Technology*, **9**, 221
7. Smith and Stewart (1979) *Metal Science*, **13**, 429

8. Barsom, R. (1971) *Proceedings of the Second NACE Conference on Corrosion Fatigue* (Connecticut, USA)
9. Bardal, E., Berge, T., Grovlen, M., Haagensen, P.T. and Forre, B.M. (1981) *International Symposium on Fatigue Thresholds,* EMAS, Warley, UK, p. 31
10. Austen, I. M. and Walker, E. F. (1987) In *Fatigue 87,* EMAS, Warley, UK
11. Radon, J. C. (1979) *Metal Science,* **13**, 411
12. Dias, A., Bignonnet, A. and Lieurade, H. P. (1987) In *Fatigue 87,* EMAS, Warley, UK, p. 749
13. Vasudevan, A. K. and Bretz, P. E. (1983) In *Fatigue Crack Growth Threshold Concepts,* TMS-AIME, USA, p. 25
14. Bignonnet, A., Namdir-Irani, R. and Truchon, M. (1982) *Scripta Metallurgica,* **16**, 795
15. Okabayashi, K., Kawamoto, S. and Nodani, H. (1966) *J. Jap. Foundryman Society,* **38**, 501
16. Bowden, F. P. and Tabor, D. (1950) *The Friction and Lubrication of Solids,* Clarendon Press, Oxford
17. Krause, H. (1971) *Wear,* **18**, 403
18. Bamford, R. (1977) In *The Influence of Environment on Fatigue.* International Mechanical Engineering Conference, London, p. 51
19. Ward-Close, M. and Beevers, C.J. (1980) *Metallurgical Transactions,* **11A**, p. 1007
20. Taylor, D. (1985) *Compendium of Fatigue Thresholds and Growth Rates,* EMAS, Warley, UK
21. Bouchet, B., Kwon, J. H., Bignonnet, A., Namdir-Irani, R. and Petit, J. *Proc. 37ieme Reunion Int. de Chemie Physique,* paper 36
22. Cooke, R. J., Irving, P. E., Booth, G. S. and Beevers, C. J. (1975) *Engineering Fracture Mechanics,* **7**, p. 69
23. Ritchie, R. O., Suresh, S. and Moss, C. M. (1980) *Journal of Engineering Materials Technology,* **102**, 293
24. Suresh, S. and Ritchie, R. O. (1982) *Metal Science,* **16**, 529
25. Suresh, S., Zamiski, G. F. and Ritchie, R. O. (1981) *Metallurgical Transactions,* **12A**, 1435
26. Suresh, S. and Ritchie, R. O. (1983) *Engineering Fracture Mechanics,* **18**, 785
27. Ritchie, R. O., Suresh, S. and Liaw, P. K. (1981) *First Ultrasonic Fatigue Conference,* (Champion, PA), AIME, USA, p. 443
28. Stewart, A. T. (1980) *Engineering Fracture Mechanics,* **13**, 463
29. Usami, S. and Shida, S. (1979) *Fatigue of Engineering Materials and Structures,* 471
30. Michel, D. J. and Smith, H. H. (1981) *Journal of Nuclear Materials,* **103**, 871
31. Salivar, G. C. and Hoeppner, D. W. (1979) *Engineering Fracture Mechanics,* **12**, 181
32. Usami, S., Fukuda, Y. and Shida, S. (1983) *Proceedings of the Fourth National Congress on Pressure Vessel Technology,* ASME, p. 1
33. Hippsley, C. A. (1987) *Materials Science and Technology,* **3**, 921
34. Smith, P. (1987) *Fatigue and Fracture of Engineering Materials and Structures,* **10**, 291
35. Petit, J. and Zeghloul, A. (1981) *International Symposium on Fatigue Thresholds,* EMAS, Warley, UK, p. 21
36. Petit, J., Bouchet, B., Gasc, C. and DeFouquet, J. (1977) *Proceedings ICF4* (Canada, 1977), Pergamon, Oxford, p. 867
37. Petit, J., Renaud, P. and Violan, P. (1982) *Proceedings ECF4,* EMAS, Warley, UK, p. 426
38. Bathias, C., Pineau, A., Pluvinage, J. and Rabbe, P. (1977) *Proceedings ICF4* (Canada, 1977), Pergamon, Oxford, p. 1283
39. Vasudevan, A. K. and Suresh, S. (1982) *Metallurgical Transactions,* **13A**, 2271
40. Suresh, S., Palmer, I. G. and Lewis, R. E. (1982) *Fatigue of Engineering Materials and Structures,* **5**, 133
41. Vasudevan, A. K., Bretz, P. E., Miller, A. C. and Suresh, S. (1984) *Materials Science and Engineering,* **64**, 113

42. Busby, A. K., Martin, J. W. and Holroyd, N. J. H. (1988) *Materials Science and Technology,* **4**, 518
43. Bailon, J-P., El Boujadaini, M. and Dickson, J. I. (1983) *Fatigue Crack Growth Threshold Concepts,* TMS-AIME, USA, p. 63
44. Hoffelner, W. (1982) *Metallurgical Transactions,* **13A**, 1245
45. McEvily, A. J. and Groeger, J. (1977) *Proceedings ICF4* (Canada, 1977), Pergamon, Oxford, p. 1293
46. Hicks, M. A., Jeal, R. H. and Beevers, C. J. (1983) *Fatigue of Engineering Materials and Structures,* **6**, 51
47. King, J. E. (1982) *Fatigue of Engineering Materials and Structures,* **5**, 177
48. Mills, W. J. and James, L. A. (1988) *International Journal of Fracture,* **10**, 33
49. Hoffelner, W. (1987) *Materials Science and Technology,* **3**, 765
50. Stanzl, S. and Ebenberger, H. M. (1983) *Fatigue Crack Growth Threshold Concepts,* TMS-AIME, USA, p. 399
51. Stanzl, S. and Tschegg, E. (1981) *Acta Metallurgica,* **29**, 21
52. Moody, N. R. and Gerberich, W. W. (1982) *Fatigue of Engineering Materials and Structures,* **5**, 57
53. Tschegg, E. and Stanzl, S. (1981) *Acta Metallurgica,* **29**, 33
54. Pook, L. P. and Beveridge, A. A. (1973) *Fatigue at Elevated Temperature,* ASTM STP520, 179
55. Pook, L. P. (1975) *Journal of Strain Analysis,* **10**, 242
56. Paris, P. C., Bucci, R. J., Wessel, E. T., Clarke, W. G. and Mager, T. R. (1972) ASTM STP513, p. 141
57. Stone, D. H. (1978) *Engineering Fracture Mechanics,* **10**, 305
58. Skelton, R. P. and Haigh, J. R. (1978) *Materials Science and Engineering,* **36**, 17
59. Shih, T. T. and Donald, J. K. (1981) *Transactions of the American Society of Engineering,* **103**, 104
60. Liaw, P. K., Hudak, S. J. and Donald, J. K. (1982) *Metallurgical Transactions,* **13A**, 1633
61. Esaklul, K. A., Gerberich, W. W. and Lucas, J. P. (1983) *International Conference on Technology and Applications of HSLA Steels* (Philadelphia, USA), p. 1
62. Poon, C. and Hoeppner, D. W. (1979) *Engineering Fracture Mechanics,* **12**, 23
63. Puskar, A. and Varkoly, L. (1986) *Fatigue and Fracture of Engineering Materials and Structures,* **9**, 143
64. Liaw, P. K., Saxena, A., Swaminathan, V. P. and Shih, T. T. (1983) *Metallurgical Transactions,* **14A**, 1631
65. Scarlin, R. B. (1976) *Metallurgical Transactions,* **7A**, 1535
66. Scarlin, R. B. (1977) *Materials Science and Engineering,* **30**, 55
67. Vincent, J. M. and Remy, L. (1981) *International Conference on Fatigue Thresholds,* EMAS, Warley, UK, p. 441
68. King, J. E. (1987) *Materials Science and Technology,* **3**, 750
69. Hicks, M. A. and King, J. E. (1983) *International Journal of Fatigue,* **5**, 67
70. Liaw, P.K. and Logsdon, W. A. (1987) *Acta Metallurgica,* **37**, 1731

7 The effect of load configuration

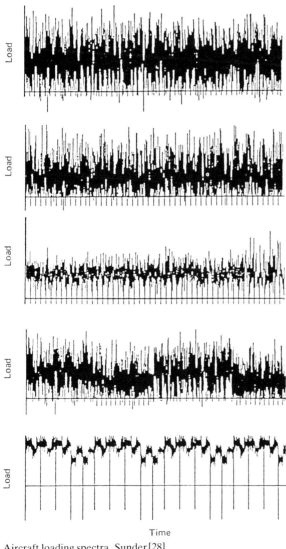

Aircraft loading spectra, Sunder [28]

Introduction

There are a variety of ways in which the type of loading experienced by a component in service may differ from that generally applied to test specimens in the laboratory. The present chapter will consider three problems:

1. The effect of *mean stress*, generally defined in terms of R ratio: There has been a great deal of data collected at R ratios of −1 and close to zero, and many studies showing the tendency for threshold to decrease as the (positive) R ratio is raised. Problem areas include the very high R ratios, approaching unity, and the high-negative and fully compressive load cycles.
2. The effect of sudden changes in load amplitude, including occasional *overloads, off-load* periods and *underloads:* a great number of studies have also been carried out on these problems, especially on the retardation effects of overloads.
3. Random loading: many structures are subject to randomly varying loads of various descriptions. A common example is narrow-band random loading resulting from resonance effects. Studies of the effects of R ratio and overload behaviour in constant-amplitude cycling are not necessarily useful to the understanding of random loading effects, as will be shown below.

The effect of *R* ratio

R-ratio effects will be considered under three headings, namely: (a) positive R ratios in the range 0–0.8; (b) very high R ratios approaching unity; and (c) negative R ratios and fully compressive loading.

Positive *R* ratios

Figure 7.1 shows a large collection of data from various workers, taken from [1], plotting the value of measured threshold as a function of R. The data have been normalised with respect to the threshold at $R = 0.5$. The plot shows a surprising regularity in the R ratio effect for this very wide range of metal alloys. For example, the scatter at $R = 0$ on this plot is plus-or-minus 19%, which is about the same as the expected accuracy of determination of threshold values themselves. At high R values results tend to scatter, with some materials showing an apparent levelling off to a constant, R-independant threshold value. However, the majority of data remain within the scatter band, even to very high values of R, approaching 1, though examination of the results of individual studies shows a tendency for data to curve down towards zero as R is increased above 0.9.

It seems reasonable, therefore, to divide the positive R ratios into two regions, namely a region of moderate positive values, which extends up to 0.8 in most materials, and a 'high-R' region above this. In the first region behaviour is basically characterized by ΔK, with an imposed R-ratio effect,

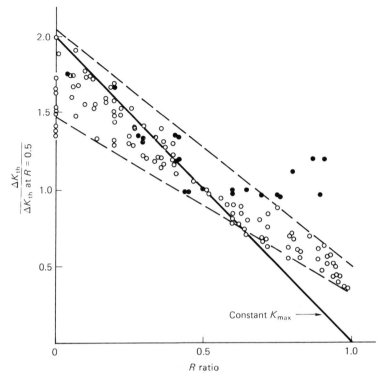

Figure 7.1 Threshold values for a variety of alloys from Taylor[1]. Data have been normalized with respect to the threshold at $R = 0.5$. Full circles indicate results which tend to show a levelling-off effect at high R

explainable by a number of mechanisms, including closure. The second, high-R region, will be discussed below.

Crack closure concepts have, on the whole, been very successful in explaining R-ratio effects (Chapter 3). The simplest closure model assumes a constant value of K_{op}, above which the crack is open, so that at low R, where K_{min} is less than K_{op}, the effective stress intensity range is given by $(K_{max} - K_{op})$. This predicts that the threshold will occur at a constant value of K_{max} and leads to a linear slope, as shown on Figure 7.1, levelling off to a horizontal line at some R ratio, $K_{min} = K_{op}$. The value of this R ratio will be expected to vary from material to material, but should rarely be higher than 0.6, considering known measurements of K_{op}.

Unfortunately, the picture is not so simple in practice. It turns out that K_{op} generally varies with R in a complex manner, often increasing as R increases. This may be the reason why the constant K_{max} line on Figure 7.1 seems to form an upper bound to the data in the range $R = 0$–0.3; points which fall on this line represent cases where K_{op} is constant with R. Experimentally measured ΔK_{eff} values are found to be roughly constant at the threshold for all R ratios, though significant increases or decreases may occur above $R = 0.5$. Figure 7.2 shows some typical plots of $\Delta K_{eff,th}$ (the value of ΔK_{eff} at the threshold) as a function of R.

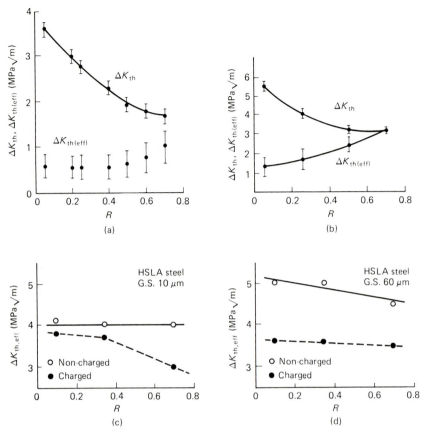

Figure 7.2 Typical results of $\Delta K_{\text{eff,th}}$ as a function of R, taken from direct closure measurements: (a,b) [26]; (c,d) [27]

An alternative approach is based on the cyclic crack-tip opening displacement (CTOD). The static value of CTOD for a given applied K is proportional to the square of the applied stress intensity:

$$CTOD \propto K^2 \qquad (7.1)$$

Thus the range of CTOD during fatigue is given by:

$$\Delta(CTOD) \propto K_{\text{max}}^2 - k_{\text{min}}^2 \qquad (7.2)$$

Now if we assume that the range of CTOD is a constant at the threshold, it can be shown that the variation of threshold with R is given by:

$$\Delta K_{\text{th}} = \Delta K_{\text{th,o}} \left(\frac{1-R}{1+R}\right)^{\frac{1}{2}} \qquad (7.3)$$

where $\Delta K_{\text{th,o}}$ is the measured value of ΔK_{th} at $R = 0$.

Figure 7.3 compares the experimental data (the scatter band from Figure 7.1) with the predictions given by equation (7.3) and by simple closure

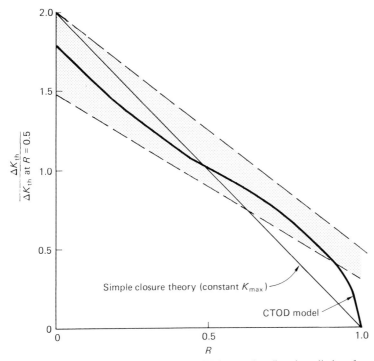

Figure 7.3 Comparison of data from Figure 7.1 (scatter band) and predictions from CTOD model (thick line) and simple closure theory (constant K_{max}, thin line)

theory (i.e. constant-K_{max}). The former gives a better approximation, especially at high R values. However, it is certain that closure effects must be acting at low R, and these are not accounted for by the CTOD model. A correction to allow for this would tend to move the CTOD prediction upwards at low R, leaving it unchanged at high R.

High R ratios

Relatively few systematic studies have been carried out on the effects of very high mean loads.

The implications of the above section are that for low R ratios, up to about 0.5, closure concepts can successfully account for the decrease in ΔK_{th}. Above this value, other considerations must come into play since closure is often non-existent[2], The cyclic CTOD model described above suggests one possible solution. Another possible mechanism which may tend to keep threshold values high in the high-R region is crack-tip blunting[3] which will reduce the degree of stress concentration at the crack tip.

Negative *R* ratios

A large amount of data has been obtained at $R = -1$, though relatively little is available at other negative R ratios, so it is difficult to comment on general trends of threshold variation in this regime. Considering the practical importance of negative R ratios (many applications involve mean compressive loads), surprisingly few workers have attempted a systematic study of their effects. Figure 7.4 compares threshold at $R = 0$ with available data on negative R values[1], for a variety of ferrous and non-ferrous alloys.

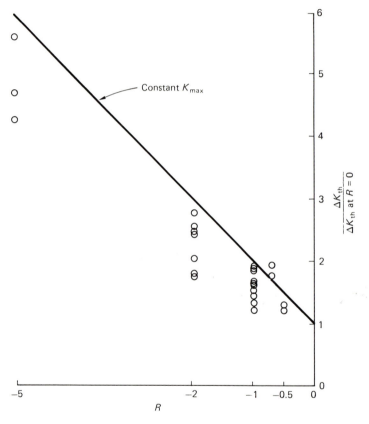

Figure 7.4 Threshold values for a variety of alloys at negative R ratios; normalized ΔK_{th}

It is common practice to assume that threshold values at $R = 0$ and $R = -1$ are simply related by a factor of two. This practice arises from the simple crack-closure assumption that the negative part of the fatigue cycle is unimportant. It is clear, however, from the diagram, that the $R = -1$ threshold is always smaller than twice the $R = 0$ value. In mild steels, for example, the ratio has an average value of 1.67, but varies widely between

1.34 and 1.9. The assumption of a factor of two is therefore non-conservative, and will overestimate the threshold by a large amount in some cases. Figure 7.4 shows that a constant-K_{max} approach such as this tends to provide an upper bound to the experimental data, but the variation in results is wide.

Mechanistically, the compressive part of the cycle must play a role; this is often referred to as crack-tip 'sharpening', but may in fact relate to changes in the plastic zone and to flattening of asperities on the crack faces. In terms of closure, this again implies a significant change in K_{op} levels, despite the fact that K_{op} will already be much higher than K_{min} at $R = 0$ (except for a very few materials).

More investigation is required into the mechanistic role of that part of the fatigue cycle which lies close to the minimum at all R ratios where closure is present. As with data at higher R values, a measured ΔK_{eff} value can be used to reconcile the results, but, as discussed in Chapter 3, the physical significance of this parameter is still somewhat unclear.

Fully compressive loading

Suresh and co-workers[4–6] have noted that cracks can be initiated from notches in specimens cycled entirely in compression. It is claimed that cracks can initiate owing to local tensile residual stresses at the notch root, set up during the unloading part of the cycle. Typically, cracks under moderate loads of 0.2–$0.4\sigma_y$ will grow for up to $0.5\,mm$ from the notch, before arresting. This appears to be due to the setting-up of closure conditions similar to those found in cracks in tension, which are sufficient to cause further growth to cease.

The effect is important in variable-amplitude loading since it implies that components in situations normally considered safe, such as far-field compression loading or compressive residual stress, may in fact initiate cracks from local stress concentrations which would then be able to grow if the component were subjected to tensile cycles.

Overloads and underloads

It has long been appreciated that overloads, i.e. load cycles whose maximum values exceed those of the normal load cycles being applied, have a strong retarding effect on subsequent crack growth. Figure 7.5 illustrates the effect; if the overload cycle is large enough, the crack growth may cease altogether for a period of cycles, and even for moderate overloads a distinct drop in da/dN will occur. This is offset to some extent, especially for multiple overloads, by higher crack growth during the overloads themselves, but generally the effect of occasional overloads is a beneficial one in that they tend to increase the overall fatigue life.

It is generally accepted that crack growth is affected for a distance which is related to the plastic zone size of the overload cycle, though whether the cyclic or monotonic zone size is to be used is unclear. The growth rate is often observed to decrease initially as the crack grows through the plastic

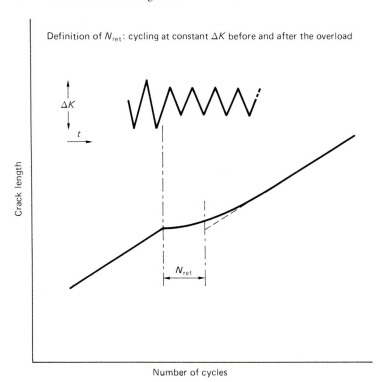

Figure 7.5 Schematic effect of an overload cycle on constant-ΔK cycling, showing retardation

zone, so that non-propagation may arise in a similar fashion to that observed in crack growth from notches (Chapter 9).

More recently it has been appreciated that underloads, i.e. cycles with relatively low minima, may also affect subsequent crack growth, causing it to increase for a period. Though the effect is not so large, and relatively large underloads are needed to cause any notable change, it is important in some applications, for instance those involving off-load periods such as aircraft components.

In the present discussion we are concerned only with near-threshold behaviour, i.e. those cases for which the overload is sufficiently severe to cause a complete or near-cessation of growth, so that threshold-like conditions have been generated. Further, the majority of examples given will be concerned with baseline stress intensity levels which are close to the conventional ΔK_{th} value, so that near-threshold growth mechanisms are relevant throughout.

Overloads

A number of models have been proposed to explain overload effects and to predict the effective value of ΔK_{th} after an overload has been applied. The models of Wheeler and Klesnil and Lucas[7] reduce to a dependence of the form:

$$\Delta K_{\mathrm{th,ol}} = \Delta K_{\mathrm{th}} \left(\frac{\Delta K_{\mathrm{ol}}}{\Delta K_{\mathrm{th}}} \right)^{\gamma} \tag{7.4}$$

where $\Delta K_{\mathrm{th,ol}}$ is the threshold produced after an application of an overload cycle ΔK_{ol}, γ being a constant which has been measured experimentally to be between 0.35 and 0.75 for various materials, and seems to be related to tensile strength[8–10].

Crack closure concepts, which are discussed in detail in Chapter 3, have been used extensively to understand overload effects. In their simplest form it is assumed that the overload establishes closure conditions, in particular a K_{op} opening value, the magnitude of which is a function of the ΔK and R ratio of the overload. It is assumed that this K_{op} value then acts on the crack after the overload, lowering ΔK_{eff}. If so, then the post-overload threshold will be given by:

$$K_{\mathrm{max}} = K_{\mathrm{op}} \tag{7.5}$$

As the crack grows, K_{op} will be modified in a manner which is difficult to predict, eventually returning to its characteristic value for the applied stress intensity. This simple approach will predict that the value of the threshold after overload, $K_{\mathrm{th,ol}}$, will depend on the intrinsic, closure-free threshold value and the value of the overload.

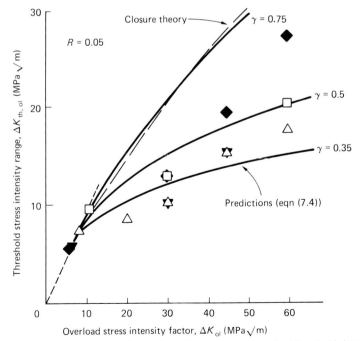

Figure 7.6 Comparison of data and predictions on post-overload thresholds in steels; data from[11]

In practice, the predictions derived from this closure model are quite similar to those resulting from equation (7.4); Figure 7.6 compares some data for steels[11], with equation (7.4) and closure predictions, assuming a threshold of 7 MPa $\sqrt{m}$ and typical closure parameters. It can be seen that the predictions from equation (7.4) are surprisingly linear for all reasonable overloads, and are relatively unaffected by the value of gamma.

The rule-of-thumb seems to be that, for overloads applied to cracks growing in the near-threshold region, the effective threshold is approximately equal to the overload value if the overload is small (say less than 2 × ΔK_{th}), otherwise it has a lower value, in the region of one-half to one-third of the overload value.

This implies that occasional overloads will dominate near-threshold behaviour; this may explain the relatively rapid attainment of threshold growth conditions for random waveforms, compared to the predictions of non-interactive models (see random loading below).

Other factors which may play a role in post-overload growth are cyclic crack opening displacement, residual stress and crack deflection. These are discussed by Alexander and Knott[12].

Underloads

Underloads may take one of three forms:

1. relatively low minima in fully tension/tension cycling
2. off-load periods at zero stress during tension tension cycling
3. high compression excursions during tension/tension or tension/compression cycling.

Underloads can cause temporary accelerations of crack growth, mirroring the retardations caused by overloads, but these accelerations are generally of smaller magnitude for comparable conditions. Considering threshold conditions, it has been shown[13] that a crack cycled at a ΔK value below ΔK_{th} can be induced to grow by applying an underload. Admittedly, the underload in this case was large – five times the normal ΔK – and the crack growth continued for a limited length of the order of 200 μm before arresting, but the implication is that periodic underloads of this kind can maintain crack growth at stress intensities below the normal threshold.

In a more comprehensive study, Yu et al.[14,15] measured thresholds and crack growth rates in the presence of regular underloads of various magnitudes, applied at varying intervals. Figure 7.7 shows some of their results, in which it is demonstrated that a compressive underload, applied as infrequently as once every 200 000 cycles, is capable of decreasing ΔK_{th} by a significant amount. The underload was, again, relatively large, having a value of 208 MPa in compression in a material of yield strength 353 MPa. Smaller, though significant effects were also noted for off-load periods imposed on high-R cycling, which would be a model for aircraft operating conditions, for example.

Figure 7.8 summarizes the effects of compression underloads on ΔK_{th} for a number of materials[14]. Generally speaking, underloads have to be of larger magnitude than overloads in order to be effective, but the

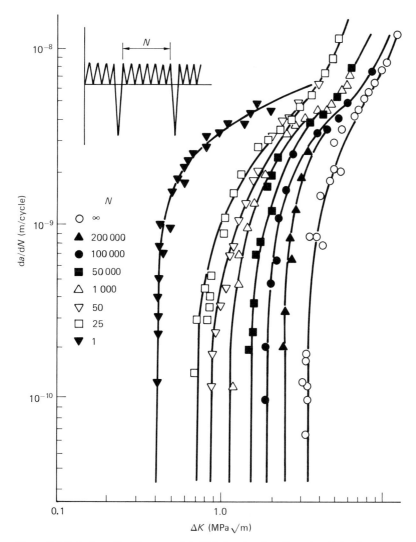

Figure 7.7 Data from Yu, Topper and Au[14] showing the effect of underloads on crack propagation behaviour

difference is not so great as might be supposed at first glance. Significant effects have been recorded for underloads at twice the normal ΔK value[16] in an Al–Li alloy; an effect of equal magnitude might typically be achieved by an overload of 1.5 times ΔK, though the magnitude of both effects would be expected to vary considerably with material, R ratio, and other factors. Overloads are generally more effective when superimposed on low-R cycles, because of the enhanced closure effects.

The mechanisms by which underloads operate have yet to be fully elucidated. Closure may be involved, since a large compressive load might be expected to cause reversed plastic flow around the crack tip, tending to

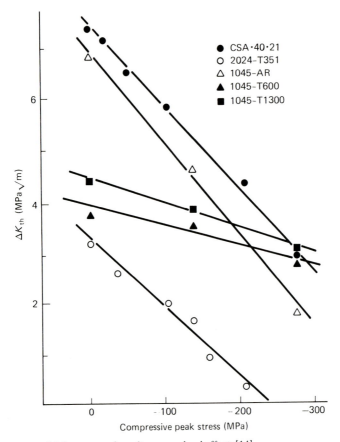

Figure 7.8 Summary of results on overload effects [14]

reduce or reverse the normal residual stress pattern. Also, severe compression might be expected to flatten crack-face asperities and therefore reduce closure for subsequent cycles. On the other hand, Suresh and Ritchie[17] have suggested that underloads may enhance oxide-induced closure by causing more fretting between crack faces.

Random loading

Compared to the two previous categories, relatively little work has been done on the effects of random load cycles. In practice, both in service situations and in testing work, the instance of 'truly' random loading, in which the value of one load peak is totally unconnected to the value of the previous peak, is rare. Probably the commonest real-life situation is that of narrow-band random loading, which arises from resonance in a system which receives random external stimuli.

Narrow-band random loading

A typical narrow-band cycle is illustrated in Figure 7.9; the important features are:

1. The frequency is almost constant
2. The rate of change of amplitude from one cycle to the next is relatively small, so that severe overload effects are not expected.

The waveform may be described by a Raleigh distribution:

$$P\left(\frac{S}{\sigma}\right) = \frac{S}{\sigma} \exp\left(\frac{-S^2}{2\sigma^2}\right) \tag{7.6}$$

where $p(S/\sigma)$ is the probability of a peak value of (S/σ), S being the stress and σ the root-mean-square (rms) value of the whole distribution. The form of the Raleigh distribution is shown in Figure 7.10. This type of loading may be simulated relatively easily by modifying a resonance-type fatigue machine[18].

Figure 7.9 Typical narrow-band random load cycle

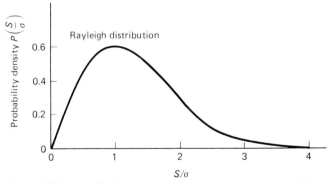

Figure 7.10 Raleigh distribution for narrow-band random loading[23]

Narrow-band spectra may be recorded from real structures, for instance offshore oil platforms. However, in such complex structures, which have more than one degree of freedom, several different resonant frequencies usually occur, effectively producing several narrow-band spectra superimposed on one another. If the various frequencies happen to be harmonics of one another, unfortunate synergistic effects may occur; for instance, the correct combination of odd harmonics will produce features similar to square waves, which may be much more damaging from a fatigue point of view.

Other types of random loading

Another important category is the type of waveform which contains occasional large deviations which are randomly distributed. An example is aircraft loading patterns (Figure 7.11) which are dominated by in-flight overloads and by stop/start cycles. Such waveforms might be approximated

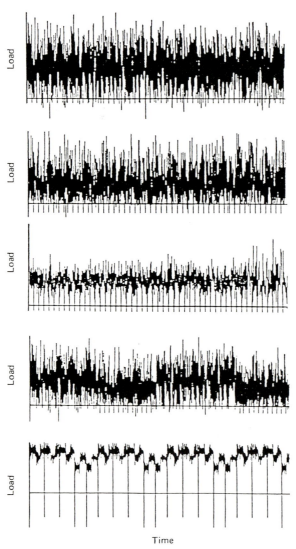

Figure 7.11 Typical aircraft loading sequences; military aircraft (fighter and transport); from Sunder[28]

to constant-amplitude cycles with periodic, randomly spaced excursions, as has been attempted by Arone[19]. If the overloads are sufficiently rare, they may constitute a Poisson distribution.

It is important to note that, in laboratory testing, two distinctly different methods are used to generate random waveforms. The first method uses some form of random signal generated electronically which is then fed to a load-producing device. Filters may be used to limit frequency and amplitude, and in some cases the random signal may be imposed on a resonant system. In the second approach, a computer program is used to generate the waveform on a cycle-by-cycle basis. In this case the individual cyclic amplitudes may be random but the rules used for their generation may include features such as a constant mean stress or the fact that each peak is followed by a trough of equal magnitude. These differences of approach may well yield different results, as shown below.

Predictive methods

Methods of analysis which have been used to predict the effect of random loading include:

1. The use of root-mean-square (rms) value of stress or stress intensity: this approach ignores any interaction effects, i.e. overload/underload effects such as were discussed in the previous section, and it also tends to underestimate the importance of the higher load amplitudes.
2. A non-interactive cumulative damage approach in which crack growth is estimated on a cycle-by-cycle basis using constant-amplitude growth data: this is effectively Miner's law adapted to crack propagation. A feature of this approach is that the threshold is less well defined because, in a random distribution of cycles, occasional high-stress cycles will exceed the threshold, even when the mean value is much lower than ΔK_{th}; this leads to a prediction of very slow growth when the growth rate is averaged over all the cycles.
3. Interactive cumulative damage approaches, in which overload and underload effects are taken into account, using approaches discussed in the preceding section.

A general criticism of any cycle-by-cycle approach is that, under near-threshold conditions, we know that crack advance does not occur on a cycle-by-cycle basis. Average crack propagation rates per cycle are less than the lattice spacing of the material, implying that portions of the crack front must lie dormant for many cycles, advancing only infrequently. This suggests that damage must accumulate ahead of the crack over tens or hundreds of cycles until conditions are appropriate for advance.

Under these conditions, the effects of very large overloads or underloads may be similar to their effects on otherwise constant-amplitude loading, but the same interaction ideas are unlikely to be appropriate for, say, narrow-band random loading such as that of Figure 7.9. Studies of the cyclic stress/strain behaviour of metals have shown that it takes a large number of cycles to establish stable cyclic stress/strain conditions under fully plastic loading; under random loading, conditions in the crack-tip plastic zone may never stabilize before crack growth occurs. It may therefore be fruitful to pay more attention to results of low-cycle fatigue testing under random loading as a key to behaviour of material in the crack-tip process zone.

Cycle counting

An additional problem in this area is the method used to count cycles and assign stress amplitudes. In the narrow-band case, each cycle can be clearly defined, and a stress range assigned to it, but in more complex random waveforms it is less clear what constitutes a single peak-to-trough cycle. Two methods commonly used are rainflow counting and spectral analysis. The latter produces stress/frequency spectra by Fourier transform methods, allowing individual frequencies to be assessed separately.

Experimental results

Relatively few experimental studies have been conducted in this area; those which are available tend to be conflicting, and suggest that, under

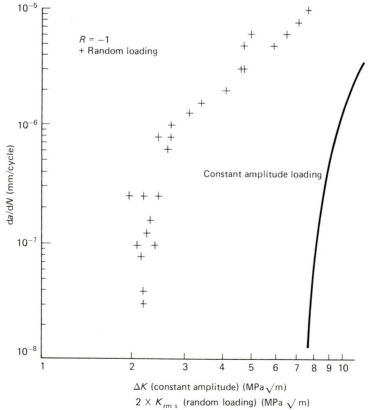

Figure 7.12 Thresholds in mild steel measured under random and constant-amplitude cycling, Kitagawa, Fukuda and Nishiyama[20]

random loading, the effects of other variables such as frequency and corrosion may be modified significantly.

Kitagawa et al.[20] and Pook and Greenan[21] have both studied fatigue of mild steel under narrow-band random loading, but have reached quite different conclusions. Kitagawa et al. demonstrated a significant decrease in ΔK_{th} under random loading (Figure 7.12); their results are expressed in terms of the rms deviation of K, but even when this is allowed for by multiplying by $\sqrt{2}$ the effective threshold is still less than half the constant amplitude value. An important feature of narrow-band random loading is the clipping ratio, defined as the ratio between the maximum peak stress and the rms value. The true Raleigh distribution (Figure 7.9) extends to infinity, but in practice there is a cut-off such that the clipping ratio rarely exceeds five. Kitagawa et al. note that for their results there is an approximate relationship:

$$\frac{(K_a)_{th}}{(K_{rms})_{th}} \approx \text{clipping ratio} \qquad (7.7)$$

where $(K_a)_{th}$ is the amplitude of K at the threshold for constant-amplitude cycles (i.e. $\Delta K_{th}/2$) and $(K_{rms})_{th}$ is the rms amplitude at the threshold for random cycles. This relationship simply implies that, for low stress cycling, the crack only grows under the occasional, high-load cycles, the threshold being reached when the largest cycles drop below the conventional threshold value.

By contrast, Pook and Greenan[21], testing quite a similar mild steel under similar conditions, observed very little change in threshold. Their K_{rms} threshold values were approximately twice those of Kitagawa et al., so it is not impossible that one or the other group of workers has miscalculated the stress intensity values. Figure 7.13 shows Pook and Greenan's data and predictive models. Here, a simple no-interaction cumulative damage approach gives a prediction of threshold which is too low, and which presumably must tend to the same prediction as Kitagawa's, expressed in equation (7.7) above, though this would only

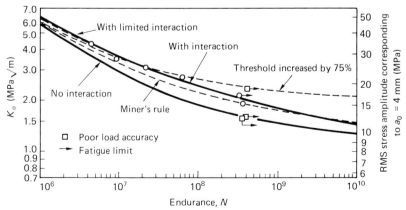

Figure 7.13 Results and predictions on random-load cycling of a mild steel, from Pook and Greenan[21]

occur at very long lives. The best fit to their data arose from incorporating an interaction law for overload cycles developed by Klesnil and Lucas (see section on overloads above); no allowance was made for underload effects, or for transient growth during an overload.

Some recent data from Austen and Walker[29] serve to complicate the picture still further. They tested two different steels in air, water and seawater at low frequencies. Thresholds tended to be higher for random loading, especially at low R ratios where increases up to a factor of 1.8 were recorded. Two features of these experiments may have played a role: first the low frequency used (which is discussed in Chapter 4 on methods of measurement), and second the method used for generating the random cycles. These were generated on a cycle-by-cycle basis, probably at constant mean stress and possibly consisting of complete cycles, so that each peak stress was followed by a trough of equal magnitude. Thus the cycles would resemble a narrow-band waveform (Figure 7.6) except that each cycle's amplitude would be unrelated to the preceding one. More pronounced interactive effects might thus be expected.

Mixed mode loading

Many practical situations involve combinations of the loading modes I, II and III, and in recent years a considerable amount of work has been done to attempt to define crack propagation in modes II and III and, more importantly, to define rules governing mixed-mode loading. These rules generally take the form of a locus in two or three dimensions, showing the combination of K_I, K_{II} and K_{III} which would be needed to achieve a particular effect (e.g. [22]).

Unfortunately relatively little near-threshold work has been carried out. Pook and Frost[23] assume that fatigue cracks cannot grow in pure mode II loading on account of the large amount of fretting action involved; this may certainly be the case at low R ratios when the crack opening is very small, effectively preventing shear movements, but mode II loading would certainly be expected to affect closure in a mixed mode I/II situation (see Chapter 3).

Experimentally, it is generally observed that cracks subjected to mixed-mode loading will grow in such a way as to maximize K_I[24], so initial threshold values might be estimated simply by calculating the K_I contribution on the initial crack[22] or, more conservatively, by calculating the K_I value for the crack if it is allowed to reorient itself to maximize K_I.

Other results, on mixed-mode I/III loading[25], are more pessimistic. It is argued that shear loading tends to flatten crack faces by removing asperities; the reduction in roughness-induced closure will therefore tend to decrease K_I down to its closure-free value, which will be much lower at low R ratios (see Chapter 3). This is substantiated by the observation that badly aligned CT specimens tend to give lower threshold values in laboratory tests owing to a mode III contribution, as discussed in Chapter 4.

A possible conservative approach to mixed-mode loading, then, is to combine the above approaches, assuming a crack orientation which gives the highest K_I value, and assuming a closure-free threshold value (e.g. approximately $3\,\mathrm{MPa}\sqrt{m}$ for an alloy steel). This approach may, however, be highly conservative for some loading patterns.

References

1. Taylor, D. (1985) *A Compendium of Fatigue Thresholds and Crack Growth Rates*, EMAS, Warley, UK
2. Powell, B. E. and Duggan, T. V. (1986) *International Journal of Fracture*, **8**, 187
3. Braid, J. E. M., Taylor, D. and Knott, J. F. (1988) *Canadian Metallurgical Quarterly*, **26**, 161
4. Suresh, S. (1985) *Engineering Fracture Mechanics*, **21**, 453
5. Christman, T. and Suresh, S. (1984) Brown University Report DE-FG02-84ER45167
6. Holm, D. K., Blom, A. F. and Suresh, S. (1985) Aeronautical Res. Inst. of Sweden Report No. FFA TN 1985–86
7. Klesnil, M. and Lucas, P. (1972) *Engineering Fracture Mechanics*, **4**, 77
8. Pook, L. P. and Greenan, A. F. (1976) In *Fatigue Testing and Design*, Vol. 2, Society of Environmental Engineers, London, p. 30
9. Pook, L. P. (1977) NEL Report No. 645, The National Engineering Laboratory, UK
10. Marci, G. (1978) *Eleventh National Symposium on Fracture Mechanics* (Virginia, USA)
11. Ogawa, T., Tokaji, K., Ochi, S. and Kobayashi, H. (1987) In *Fatigue 87*, EMAS, Warley, UK, p. 869
12. Alexander, D. J. and Knott, J. F. (1987) In *Fatigue 87*, EMAS, Warley, UK, p. 395
13. Ritchie, R. O., Zaiken, E. and Blom, A. F. (1984) In *Fundamental Questions and Critical Experiments on Fatigue*, ASM, USA
14. Yu, M. T., Topper, T. H. and Au, P. (1984) In *Fatigue 84*, EMAS, Warley, UK, p. 179
15. Yu, M. T., Topper, T. H., Du Quesnay, D. L. and Levin, M. S. (1986) *International Journal of Fracture*, **8**, 9
16. Yu, W. and Ritchie, R. O. (1987) *Transactions of the American Society of Mechanical Engineering, Engineering Materials and Technology*, **109**, 81
17. Suresh, S. and Ritchie, R. O. (1987) In *Fatigue Crack Growth Threshold Concepts*, TMS-AIME, USA, p. 227
18. Very high cycle fatigue, *Proceedings of the Conference on Fatigue of Welded Structures*, The Welding Institute, Cambridge, p. 273
19. Arone, R. (1988) In *Fatigue 87*, EMAS, Warley, UK, p. 407
20. Kitagawa, H., Fukuda, S. and Nishiyama, A. (1978) *Bulletin of the Japan Society of Mechanical Engineers*, **21**, 367
21. Pook, L. P. and Greenan, A. F. (1979) *International Journal of Fatigue*, **1**, 17
22. Otsuko, A., Tohgo, K. and Matsuyama, H. (1987) *Engineering Fracture Mechanics*, **28**, 721
23. Pook, L. P. and Frost, N. E. (1973) *International Journal of Fracture*, **9**, 53
24. Pook, L. P. and Greenan, A. F. (1978) NEL Report No. 654, Department of Industry, UK
25. Nix, K. J. and Lindley, T. C. (1988) *Second International Parsons Turbine Conference* (PITC2) (Cambridge, 1988) Institute of Metals, UK
26. Blom, A. F. (1984) In *Fatigue Crack Growth Threshold Concepts*, TMS-AIME, USA, p. 263
27. Esaklul, K. A., Wright, A. G. and Gerberich, W. W. (1984) In *Fatigue Crack Growth Threshold Concepts*, TMS-AIME, USA, p. 299
28. Sunder, R. (1988) In *Fatigue 87*, EMAS, Warley, UK, p. 185
29. Austen, I. M. and Walker, E. F. (1988) In *Fatigue 87*, EMAS, Warley, UK, p.1155

8 Short cracks

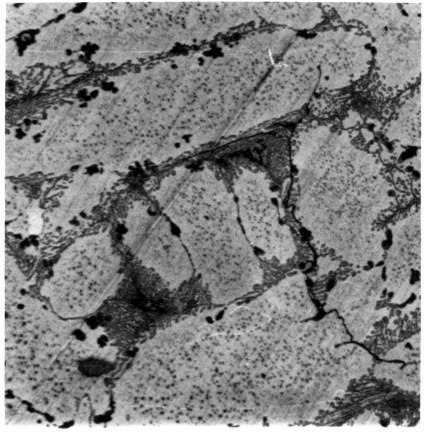

Short crack and microstructure – nickel aluminium bronze alloy

Introduction

Improvements in our ability to detect and measure very small cracks, combined with a greater interest in using the methods of fracture mechanics for smaller and smaller flaw sizes, has, in recent years, highlighted a problem. It has been shown that very small cracks do not obey the same propagation laws which we have been able to apply to long cracks; specifically, the standard dependence of crack propagation rate, da/dN, on stress intensity range, ΔK, no longer applies if the crack is less than some critical length, typically 1 mm in a mild steel or low-strength alloy, and decreasing to as little as 10 μm in some high-strength materials.

It is clearly important to understand this phenomenon, since it puts a lower limit on our use of standard threshold data for assessment of defect tolerance. It is also necessary to have some understanding of the various explanations which have been put forward for this behaviour, and to develop some rational approach for assessment of defects in the size range which may be present in engineering components.

The amount of research work and data which has appeared on this subject in recent years has tended to mask the fact that most practical applications of materials involve relatively large pre-existing flaws, so that the short-crack problem only arises in a limited number of real applications. These, however, include some important, high-risk areas such as aeroengine components and some high-volume applications such as well-polished surfaces on automotive components. These and other examples will be discussed towards the end of this chapter, which begins with a basic outline of the phenomenon and its mechanistic explanations.

Two excellent sources of background reading in this area are Suresh and Ritchie's 1984 review[1], which was the first comprehensive review of work in this area, and the proceedings of the EGF conference, *The Behaviour of Short Fatigue Cracks*[2] published in 1986.

Short-crack behaviour: the basics

It is wise to begin with some definition of terms. In the following, as in most other texts, the term 'short crack' will refer specifically to a crack which demonstrates the anomalous behaviour described below. The terms 'small crack' and 'physically short crack' are used to refer to cracks which have lengths in the millimetre range or less, but which do not necessarily behave anomalously. Finally, the term 'microstructurally short crack' refers to a crack which is small compared to the predominant microstructural features of the material; for instance in a mild steel this would be a crack equal to, or smaller than, a few grain diameters in length.

It is worth remembering that a crack subjected to constant cyclic stress conditions will increase its length as shown schematically in Figure 8.1, spending the great majority of its time, and therefore of the lifetime of the component, at a length very close to its original length. As a rough guide it is fair to say that by the time the crack has increased its length by a factor of 10, most of the life of the component is over. Thus it is of vital importance

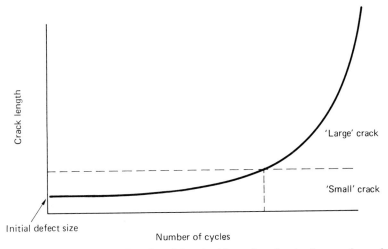

Figure 8.1 Schematic variation of crack length with number of cycles for a crack growing under constant applied stress Δσ

to understand the growth behaviour of small cracks in all cases where these may be the largest available defects in the material.

The anomalous behaviour observed in short cracks is illustrated schematically in Figure 8.2. Here a typical curve derived from testing long cracks is compared to results from *one* short crack, keeping Δσ constant, so

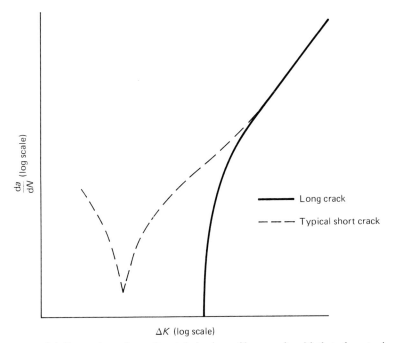

Figure 8.2 Comparison of growth-rate behaviour of long cracks with that of one typical short crack, tested at constant Δσ

that the increase in ΔK is caused by the increase in crack length. Compared to the usual long-crack behaviour, three features can be noted:

1. Cracks grow at higher da/dN values than would be predicted by the long-crack curve, for a given ΔK.
2. Cracks demonstrate growth at ΔK values lower than the long-crack threshold.
3. An individual short crack often shows a minimum in da/dN with increasing ΔK.

In retrospect this behaviour is not as surprising as it first appears. Consider a plot such as Figure 8.3, in which the value of stress range at the threshold, $\Delta\sigma_{th}$, is plotted as a function of crack length, using logarithmic scales. If ΔK_{th} is a material constant, the line marked 'threshold', with a slope of -0.5, should apply. This would predict an ever-increasing value of $\Delta\sigma_{th}$ if crack length is decreased. However, we know that if the crack length is zero, for a perfectly polished specimen, the threshold stress for fatigue is not infinity, but is equal to the fatigue limit, $\Delta\sigma_0$. So we can draw a horizontal line at this value of stress on Figure 8.3. This implies that for

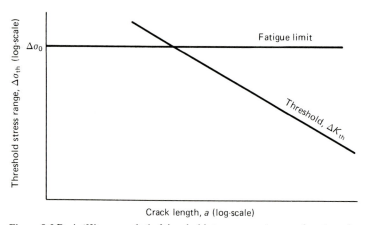

Figure 8.3 Basic 'Kitagawa plot' of threshold stress range $\Delta\sigma_0$ as a function of crack length

small cracks, the threshold stress range will be less than that predicted by the ΔK_{th} value; such cracks will be observed to grow at applied ΔK values less than ΔK_{th}, and therefore display the behaviour shown in Figure 8.2. The method of representation shown in Figure 8.3 is often called a 'Kitagawa' plot after one of its originators.

The experimentally derived variation of threshold stress with crack length takes the form shown in Figure 8.4; the measured threshold deviates from the long-crack prediction at some value of crack length, and eventually merges with the fatigue limit. It is important to note that in the curved region of the plot, the threshold stress is *less* than would be predicted from either the fatigue limit or the long-crack threshold. Simple predictions would thus be non-conservative whether an *S/N* curve approach or an LEFM approach is used.

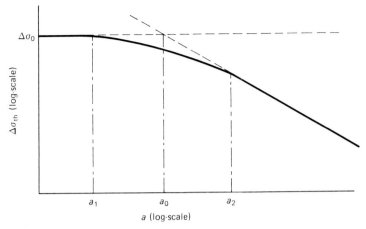

Figure 8.4 Typical experimental behaviour of short cracks, plotted on the Kitagawa diagram

Three crack-length parameters can be defined from the plot. It will be found that different researchers have used a variety of symbols for these parameters, but the following are now becoming generally accepted:

1. a_0 the crack length at which the two straight-line predictions cross
2. a_1 the length at which behaviour deviates from fatigue limit behaviour; i.e. the smallest crack which is able to lower the fatigue strength
3. a_2 the length at which behaviour deviates from long-crack threshold behaviour; i.e. the 'critical crack length' at which short-crack behaviour begins.

Scatter in short-crack behaviour

If the behaviour shown schematically in the above figures was uniform and reproducible, short-crack behaviour would be relatively easy to deal with. Unfortunately, each short crack shows different behaviour, as represented by Figure 8.5. Note that:

1. Some cracks show the decreasing part of the curve only, eventually arresting and therefore showing a kind of threshold behaviour.
2. Data from all the cracks tend to merge with the long-crack data as crack lengths approach $a2$.

This implies that similar scatter must be seen in the Kitagawa plot, as shown in Figure 8.6. Note that the lines curve downwards at very low crack lengths, and that some lines cross the fatigue limit stress. In fact, the fatigue limit stress will be the *lowest* possible value for the maxima in these curves, since fatigue failure of the specimen will occur from the crack which grows most easily. Results reported in the literature on Kitagawa diagrams tend to take the form illustrated in Figure 8.4, rather than in Figure 8.6, because the cracks observed tend to be the worst ones in the specimen, i.e. the ones which are growing most rapidly.

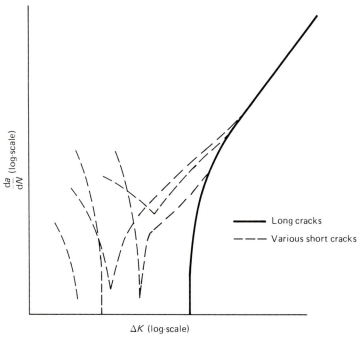

Figure 8.5 Typical behaviour of a number of different short cracks in the same material, at the same $\Delta\sigma$

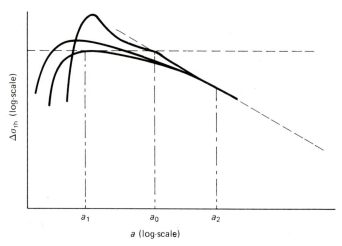

Figure 8.6 Expected behaviour of a number of different short cracks on the Kitagawa diagram

Explanations of short-crack behaviour

The basic phenomenon can be explained as follows. First, we know that, given a certain stressed volume of material, cracks will initiate from the most favourable sites. These may be geometrical, such as surface imperfections, or microstructural, such as favourably oriented grains or brittle precipitates. In all cases the local environment of the initiated crack is unusually encouraging of growth, more so than the bulk of the material. Thus it is clear that short cracks will grow more rapidly at the early stages. A simple example of this is a crack growing within one grain, which has not yet reached a grain boundary. As growth continues, the crack is forced to advance into 'normal' material, where conditions are less favourable for growth, and its increasing size forces it to sample the material more uniformly. It thus becomes a normal long crack, whose behaviour is representative of the bulk properties of the material.

Any one short crack will face unique local conditions; for instance, a crack growing in one grain will have a growth rate which is influenced by lattice orientation and the presence of grain boundaries and precipitates. Thus we must expect that each crack will behave uniquely, giving rise to the scatter described above. The minima in growth (Figures 8.2 and 8.5) are associated with the worst conditions for growth; at this stage the crack is still quite short, and therefore the applied stress intensity is low, but it is long enough to begin to encounter growth obstacles such as grain boundaries.

Several mechanisms have been advanced to rationalize short-crack behaviour, and all the following can be expected to exert an influence. As yet, the relative importance of these different mechanisms is poorly understood.

1. Microstructural effects

Owing to the sampling effect mentioned above, it is expected that a crack will not behave as a long crack until it is relatively large compared to the microstructure. For a simple single-phase material the grain boundary will be the predominant feature, so short-crack behaviour is expected to continue until the crack is at least an order of magnitude larger than the average grain size[3]. For more complex microstructures, the effective grain size defined by Yoder et al. (see Chapter 5) will be appropriate. This is defined as the average mean free path through any significant microstructural features; so, for instance, in a material dominated by precipitates this might be the average precipitate spacing.

The mechanism of rapid crack growth in this case may simply be the selection of weak microstructural paths, and decreases in crack growth may be associated with microstructural barriers. A very common observation in, for example, mild steels and aluminium alloys, is that the minimum in the growth-rate curve (Figure 8.2) is associated with the arrival of the crack front at the first grain boundary, previous growth having occurred entirely within one grain. In some cases a series of minima occur, as successive grain boundaries are encountered (Figure 8.7, from experimental data[4]). However, microstructural effects also influence other mechanisms such as closure and crack deflection, as shown below.

2. Crack closure

As discussed in Chapter 3 and subsequent chapters, most long fatigue cracks tested at low and medium R ratios, display considerable amounts of

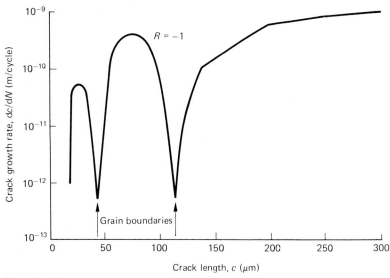

Figure 8.7 Growth minima associated with grain boundaries[4].

closure during the fatigue cycle. It can be demonstrated that short cracks show much less closure; in many cases the initial growth of a short-crack will be closure-free. This is shown in Figure 8.8, which shows threshold measurements from the same material as used in Figure 8.7[4]. Here the measured value of ΔK_{th} is equal to the closure-free value, $\Delta K_{eff,th}$, at low crack lengths, and gradually rises up to the normal long-crack value as length increases. Reduced closure levels have been confirmed by direct measurements on small cracks[18]; it has also been noted that threshold values for short cracks are similar to those for long cracks at high R values

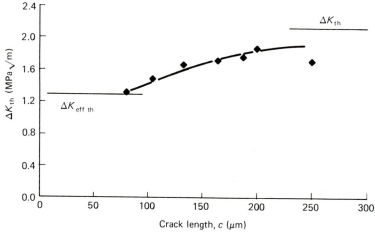

Figure 8.8 Threshold measurements from cracks in the same material as Figure 8.7, showing the effect of crack length on threshold[4]

(which are also expected to be closure-free), though this is not always the case[5].

Why should closure develop gradually in this way? This depends on the operative closure mechanism (see Chapter 3). If the residual stress in the plastic wake dominates closure, then a significant wake can only be developed after a significant amount of crack growth, presumably of the order of the plastic zone size. If, on the other hand, roughness-induced closure is dominant, the characteristic roughness will only develop after the crack has crossed a number of grains, or other microstructural features which cause crack deflection. Thus the origin of this effect may be either mechanical, dominated by plastic zone size, or microstructural, dominated by grain size.

3. High applied stress

Given a small crack length, one must necessarily impose a relatively high stress for a given ΔK value. If the stress becomes too high, normal LEFM conditions will break down, and the effect will be a relatively large plastic zone (or even general yielding) and easier crack propagation. It is not entirely clear at what value of stress the breakdown occurs; it will certainly be affected by the materials' cyclic stress/strain characteristics, but one-third of the cyclic yield strength is a good general guide. It has been noted that critical crack length ($a2$) values for some materials occur at stress levels of around this value. Lankford and Hudak[6] have noted

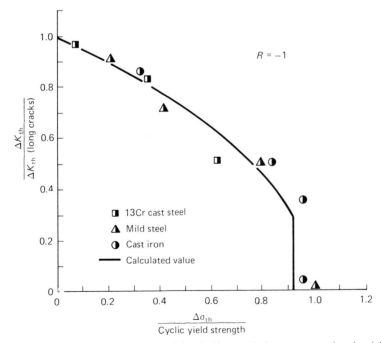

Figure 8.9 The decrease in measured threshold as applied stress approaches the yield strength, from Usami[7]

larger crack-tip strain values for short cracks at a given ΔK; microstructural effects may also cause this effect, however Holm and Blom[7] have predicted, using an elastic-plastic finite element model, that crack-tip plastic zone size will be considerably larger in a short crack, even in the absence of microstructural effects.

Figure 8.9 shows a successful prediction by Usami[8] of the decrease in threshold as the applied stress approaches the cyclic yield strength. The plot, in the form of a normalized two-parameter fit similar to a Goodman diagram, is based on plastic zone size calculations using the Dugdale formula, and predicts deviations from the long-crack threshold, even at modest applied stresses. This effect, then, is not connected with crack length as such, but only with applied stress level. In contrast to this, results on a copper alloy[3] showed consistent long-crack threshold values for cracks, even when the applied stress exceeded the monotonic yield strength. This was presumably due to cyclic hardening effects.

Calculations based on ΔJ-integral computation have also been used to rationalize short-crack behaviour under high stress fields[9].

4. Stress field effects

Even under low applied stresses, the form of the elastic singularity at the crack tip is not quite the same for short cracks as it is for long cracks[10]. This arises from the fact that the normal formulation for stress intensity is based on a simplified expression for the crack-tip stress singularity, normally written in the form:

$$\sigma = \sigma_0 \sqrt{\left(\frac{a}{2r}\right)} \qquad (8.1)$$

where σ is the stress at a distance r from the crack tip, σ_0 being the applied stress. This equation is valid only if crack length, $a >> r$; otherwise the complete form of the Westergaard equation should be used:

$$\sigma = \frac{\sigma_0[1 + (r/a)]}{[2(r/a) + (r/a)^2]^{1/2}} \qquad (8.2)$$

For short cracks, one may be concerned with events which are occurring at a distance r which is of the same order of size as the crack length itself. In such cases equation[2] should be used in full; unfortunately this does not lead to a simple definition for K. Allen and Sinclair[10], using an analysis based on plastic zone size estimation, propose the formula:

$$K_{eff} = K(1 + r_p/a)^{1/2} \qquad (8.3)$$

where K_{eff} is the effective K value of the short crack (not to be confused with the parameter ΔK_{eff} derived from crack closure) and r_p is the plastic zone size. This implies that ΔK will be larger than expected, and therefore crack growth will be faster than expected, if the plastic zone size is a significant proportion of the crack length. Clearly, this is dependent on both crack length and stress level, and in practice the predictions from this model turn out to be similar to those in the 'high stress' category above.

5. Crack deflection

The initial growth of a crack tends to be relatively planar, especially if this is confined to one grain. This used to be referred to as 'Stage I' growth, and classically produced a crack at 45° to the applied stress axis, due to shear-stress control. As the crack grows, its general orientation soon becomes normal to the tensile stress axis under simple loading conditions, but on a microscopic level the crack shows many deflections associated with individual microstructural features. This type of growth, described in Chapter 5, gives rise to smaller effective ΔK values owing to both closure and crack-deflection effects. Very short cracks, if they are free from such deflections, may show less closure and higher effective ΔK values. This problem has been analysed by Suresh[11].

6. Plane-stress effects

It should not be forgotten that a surface crack is growing in a region of plane stress, due to surface relaxations. Consider a typical semielliptic surface crack (Figure 8.10) loaded by a crack-opening σ_1 stress. Where the

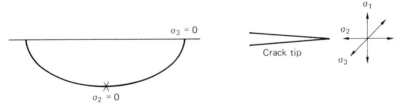

Figure 8.10 Schematic semi-elliptic surface crack and stress-axis geometry

crack front emerges from the surface, the crack will be under normal plane-stress conditions similar to those in a very thin plate specimen, since the through-thickness stress σ_3 will have a value of zero. At the point X on the crack front, plane-stress conditions will exist if the crack length is small enough, but in this case it will be the σ_2 stress (see Figure 8.10) which takes the value zero, with plane stress occurring in the σ_1/σ_3 plane. The consequences of this are not clear; however in the case of long cracks, plane-stress regions tend to show slower growth and higher thresholds than plane-strain regions (see Chapter 3) so this effect should tend to slow down the short cracks rather than to speed them up. Like the closure effect mentioned above, this will occur if the crack length is of the same order of magnitude as the plastic zone size, or smaller.

Critical length parameters: a_0, a_1 and a_2

It is useful to consider in some more detail the three length parameters defined by the Kitagawa plot, considering their physical significance, theoretical prediction and potential usefulness.

a_0: *the intersection point*

This parameter was originally defined by ElHaddad *et al.* [12], who termed it l_0. In order to model the curved part of the Kitagawa plot they used an equation of the form:

$$\Delta K_{th} = \Delta \sigma_{th} \sqrt{[\pi(a + a_0)]} \tag{8.4}$$

in which the standard equation for ΔK is modified by adding a constant, a_0, to the crack length. Figure 8.11 shows the form of this curve for a typical alloy steel. The curve fulfils the basic requirements in that it tends to a constant value at very low crack lengths, equal to the fatigue limit, and it merges with the long-crack line once crack length becomes large compared to a_0.

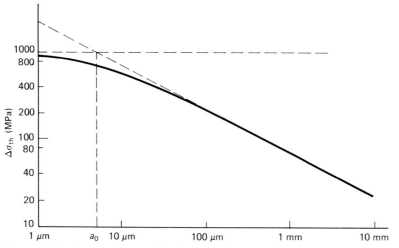

Figure 8.11 Prediction of short-crack behaviour using the ElHaddad approach, equation (8.4)

This approach suggests a physical meaning for a_0; it is implied that a crack of any given length will be able to extend by an extra amount, and at low crack lengths this concept may be appropriate, but the value of a_0 itself does not correspond to any physical size within the material, either of microstructure or of plasticity.

The usefullness of a_0 is that it can be calculated without conducting direct measurements of short cracks, using only the fatigue limit and ΔK_{th} for the material. It provides a useful initial assessment of the order of magnitude of crack length for which short-crack behaviour is likely to be a problem.

a_2: *the short/long transition*

This parameter is often referred to as the 'critical crack length', since it defines the point below which short-crack concepts will have to be used, signalling the end of LEFM applicability. For practical applications it is very important to be able to estimate a_2; if the initial defect size involved is greater than a_2 the normal techniques described elsewhere in this book may be applied, and if the defect size is slightly less than a_2 these same techniques may be modified to give a conservative estimate of behaviour, as described below.

Considering the various explanations for short-crack behaviour which are discussed above, it may be noted that these fall into two categories:

(a) *microstructural* effects, such as crack deflection, roughness-induced closure and interaction with grain boundaries.
(b) *plasticity* effects such as plasticity-induced closure, high stress and stress field problems.

Considerations under (a) might be summarized by saying that, for long-crack behaviour to occur, the crack length must be large compared to the microstructural features of importance, e.g. the grain size. Effects under (b) amount to the requirement that the crack be large compared to the plastic zone size ahead of it.

The present author has therefore proposed[3,13] that the value of a_2 may be predicted as the *largest* of the following two values:

$$a_2 = 10d \tag{8.5a}$$

$$a_2 = 10r_p \tag{8.5b}$$

where d is taken to be the average grain size or other relevant microstructural distance, as defined by Yoder *et al.* (see Chapter 5), and r_p is the cyclic plastic zone size at the applied stress level.

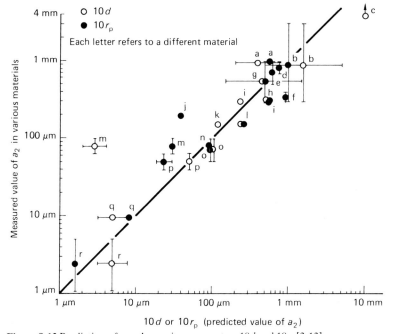

Figure 8.12 Prediction of a_2 values using parameters $10d$ and $10r_p$[3,13]

The value of 10 in these equations is arbitrary, and was simply intended to represent an order of magnitude. The equations are not expected to be exact but, as shown in Figure 8.12, they provide a good approximation to known data for a_2. In this diagram both $10d$ and $10r_p$ have been plotted; in most cases the two values are similar, but occasionally either $10d$ or $10r_p$ is significantly the larger, indicating a dominance of microstructural control or of plasticity control respectively.

Other workers have predicted short-crack behaviour based on stress level alone[8], using a Dugdale formulation for plastic zone size to assess the point at which excessive plasticity invalidates LEFM. This gives approximately the same predictions as equation (8.5b) here, but may underestimate a_2 in cases where equation (8.5a) gives the larger value, indicating microstructural control.

Some specialized materials have extremely large a_2 values; for instance, engine turbine blades made from nickel-base alloys may be in the form of single crystals or very large grain size materials, such that the value of a_2 is larger than the component itself. At the other end of the scale, high-strength alloy steels may have a_2 values of the order of $10\,\mu m$, giving short-crack behaviour only from very small defects such as surface roughness and inclusions.

a_1: the non-damaging crack

The parameter a_1 is the most difficult to measure or predict, and its significance is also debatable. One definition is that a_1 represents the largest crack whose presence does not alter the fatigue limit of the material. It has been noted in a number of coarse two-phase microstructures such as mild steels and dual-phase steels[14] that cycling for long periods at a stress slightly *below* the fatigue limit gives rise to numerous small cracks in the microstructure. It is argued that such cracks initiate very easily but become arrested as they grow, for instance by encountering microstructural barriers. The largest of these cracks should have a length of a_1. This has led some workers to propose that crack initiation is a very easy process in all materials, the critical stage being the growth of cracks beyond a_1 in length. Considering initially defect-free material, if a crack of length a_1 arises and can be induced to continue growing, then it should not arrest at a greater length, according to the Kitagawa diagram. The modified form of the diagram (Figure 8.6) illustrates this maximum value in the threshold stress level for a given crack.

Therefore the value of a_1, and the behaviour of cracks of this length, are, in theory, of paramount importance in the prediction of high-cycle fatigue behaviour, and a number of workers have proposed detailed models whose aims are to predict this behaviour[15], some of which will be discussed below. Others argue that, despite our knowledge of the crack growth mechanism that operates at these crack lengths, a design approach based on fatigue limit measurements and *S/N* curves will always be a more accurate predictor, given the inherant variability of the behaviour of cracks of this length.

Descriptions of short-crack behaviour

Figure 8.13 shows typical data on short crack propagation for a variety of materials[16,17,19,27]. The common features of these curves are a decrease to a minimum growth rate and a gradual merging with the long-crack curve. Experiments are generally carried out at near-threshold stress intensities; results at higher stress levels show similar features but their analysis is often complicated by plasticity considerations.

Miller *et al.*[18] have proposed that it is more appropriate to use a horizontal axis of crack length rather than ΔK, since the essential features of the data are crack-length controlled. In most materials the minimum in the curve corresponds to the first grain boundary, so that, assuming a random initiation point for the crack, these minima should be scattered about a mean value equal to half the average grain size, $d/2$. These workers have developed a theory in which growth rate is assumed to be related to the distance between the crack tip and the grain boundary. Figure 8.14 shows a comparison of data and prediction for a mild steel; the prediction is of the correct form, but in this case the value of d was chosen retrospectively to give the best fit to the data. It was noted that this value of d corresponded approximately to the spacing of pearlite colonies in this case.

A number of other workers have proposed similar approaches[19–21], but these vary in important details. For instance, Lankford and Davidson[19] propose that the effective plastic zone size is equal to the distance from the crack tip to the first grain boundary and that crack growth rate is related to plastic zone size, causing the crack to slow down as the boundary is approached. Cox and Morris[20], on the other hand, contend that the crack growth rate *increases* as the crack approaches the boundary. Thus both workers predict that the crack will treat the boundary as a barrier to growth, but for different reasons. The experimental evidence seems to be in favour of Lankford's and Miller's approaches in this case.

A statistical approach

Scatter is an essential feature of short-crack data, however it is presented. This arises from microstructural effects, e.g. the distance to the first grain boundary will vary from crack-to-crack, because of variation in (a) crack initiation point and (b) grain size. Scatter also arises from variations in other features of the local environment, such as residual stresses and the location of other cracks.

Considering the threshold condition only, it may be said that a crack of a given length, loaded with a given cyclic stress, has a certain *probability* of being able to grow; this probability will vary with applied ΔK and with crack length. For instance, a crack which is longer than a_2 will have a very simple probability function, shown in Figure 8.15; since growth depends only on exceeding the single-valued parameter ΔK_{th}, the probability, P, of growth, will change from 0 to 1 at a crack length of a_{th} which depends only on the applied $\Delta \sigma$:

$$a_{th} = \left(\frac{\Delta K_{th}}{\Delta \sigma} \right)^2 \frac{1}{\pi} \qquad (8.6)$$

for a simple through-crack, for instance.

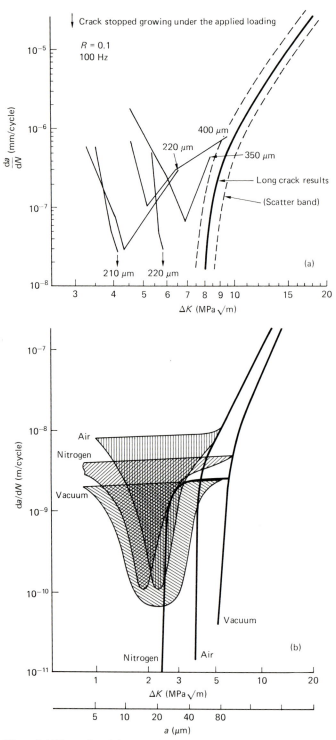

Figure 8.13 Examples of short-crack behaviour in various materials: (a) aluminium bronze [16]; (b) 7075-T651 aluminium alloy [19]; (c) aluminium alloy [17]; (d) mild steel [27]

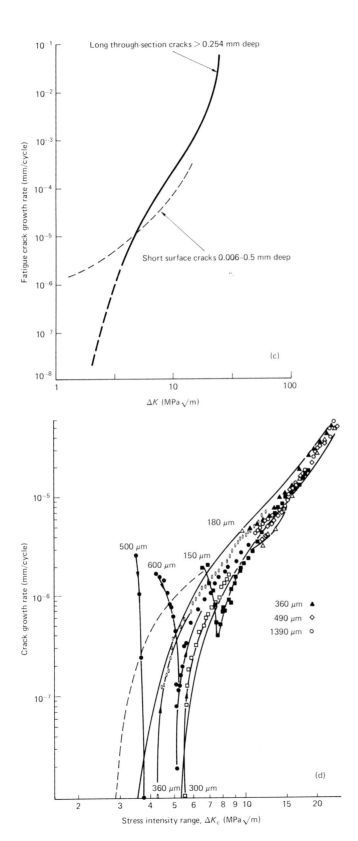

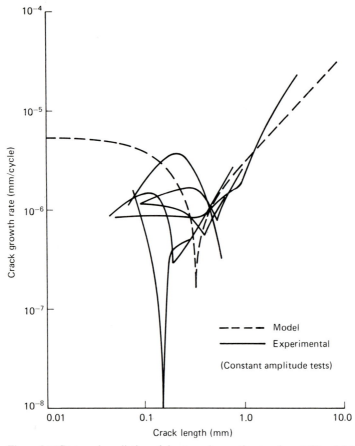

Figure 8.14 Data and prediction of short-crack growth rates, from Miller, Mohamed and de los Rios[18]

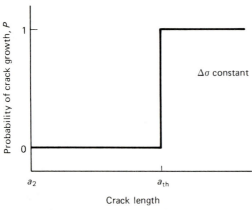

Figure 8.15 Probability of crack growth, P, as a function of crack length; simple case of a crack which is longer than a_2

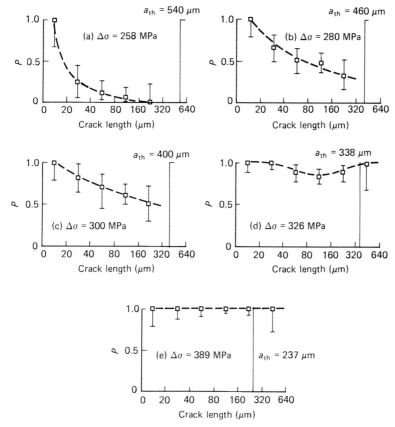

Figure 8.16 Probability, P, as a function of crack length; experimental results from a copper alloy [22] at five different applied stress levels: (a) 258 MPa; (b) 280 MPa; (c) 300 MPa; (d) 326 MPa; (e) 389 MPa

Since short cracks tend to pass through a minimum in growth rate (Figure 8.13, 8.14), this implies a minimum in the probability function, occurring at a crack length of the order of the average grain size. Figure 8.16(a–e) shows data for a large number of cracks observed at five different stress levels in a copper alloy [22], Figure 8.17 summarizes this data. For stress ranges less than or equal to 300 MPa, probability decreases sufficiently sharply with increasing crack length as to virtually prohibit continued growth; the cracks will all arrest at microstructural barriers. The stress of 326 MPa gives interesting results; in this case the minimum point is at a sufficiently high probability to ensure failure of the structure in some cases.

The failure of a component or structure made from this material will depend also on the *number* of cracks available, which is related to the stressed volume. It is possible, therefore, to calculate a probability of structure failure, Q, which incorporates P and the stressed volume. The condition for failure in this case is that a crack will grow to a length a_2,

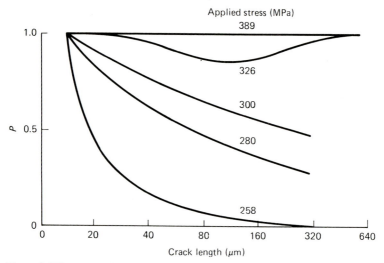

Figure 8.17 Summary of results from Figure 8.16a–e

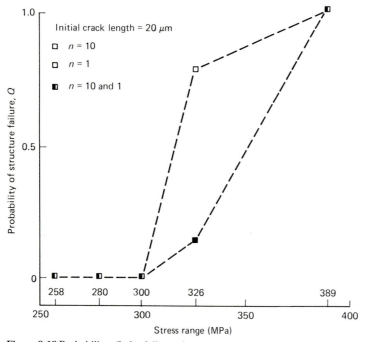

Figure 8.18 Probability, Q, for failure of a structure made from the material described in figs 8.16, 8.17; n is the number of cracks present

beyond which continued propagation is assured. Figure 8.18 shows predictions for Q, assuming one crack or ten cracks in the structure. This highlights the difference in predictions at a stress of 326 MPa, though in any real structure the number of cracks would be much larger, pushing the fatigue limit to a value between 300 and 326 MPa. Also, because the prediction is a statistical one, its degree of accuracy will increase as the number of cracks increases.

There are a number of advantages to this approach to short-crack behaviour. First, the predictions do not depend on a mechanistic model for short-crack growth, but only on the availability of a large amount of growth data. Second, the result provides a probability of structure failure which should be related to scatter on S/N curves. This can be incorporated into a wider statistical analysis of failure as used by designers, which would take into account factors such as failsafe redundancy and acceptable risks. Third, the model calculation can include a probability function for the size of the initial defect. In the above calculations a very small initial crack length of 20 μm was assumed; here the choice of 20 μm is not critical, since the value of P around 20 μm is close to 1. Knott[23] assumed a Weibull distribution of inherent defects in order to predict scatter in S/N data; in his case long-crack behaviour was assumed throughout, but the same approach could be introduced into the above analysis.

A similar probabilistic analysis can be applied to the results of other workers; a re-analysis of data provided by Elsender, Gallimore and Poynton[24] on fatigue crack growth from inclusions in a rotor steel showed P values which varied from 0.3 to 1 as inclusion size increased. The inclusions in this case were fairly large, however, so the probability of growth may be influenced by other factors such as local residual stress. Despite this, the analysis should still be valid when it comes to determining probabilities of structure failure.

Material differences

Figures 8.19–8.21, taken from a comprehensive publication by Usami[8], illustrate some typical Kitagawa plots for a variety of materials and environmental conditions. Figure 8.19 shows the effect of strength level in ferrous alloys; note also that much of the data is taken from surface defects and surface roughness, a point which is discussed in Chapter 9. Figure 8.20 illustrates the effect of temperature on a stainless steel and Figure 8.21 includes the effects of both R ratio and an aqueous environment. Data on short cracks under different environmental conditions and different R ratios are relatively scarce.

The relative importance of short-crack behaviour in different alloys is well illustrated by comparing high- and low-strength steels. Figure 8.22 compares a typical alloy steel of yield strength 1500 MPa, threshold 4 MPa √m at $R = 0$ and $a_2 = 10$ μm, and a typical mild steel with yield strength 300 MPa, threshold 9 MPa √m at $R = 0$ and $a_2 = 1$ mm. The fatigue limit is assumed to be $0.67\sigma_y$ in each case.

The first point to note is that the high-strength steel has inferior properties above a certain crack length, owing to its lower threshold; so the low-strength material is to be preferred if defects larger than 100 μm are present. The second point of interest is the stress value at which a_2 occurs, which is 713 MPa and 160 MPa for the high- and low-strength materials respectively. From a design point of view, the gap between this stress and

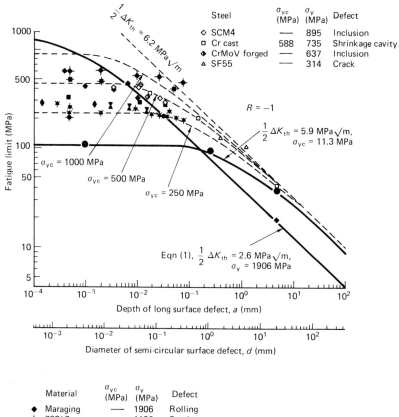

Figure 8.19 Data on various steels, from Usami [8]; initial defects include surface roughness, inclusions and casting defects as well as cracks

the fatigue limit stress represents the 'region of uncertainty', so to speak, in which short-crack concepts may have to be used. It is interesting to note that, taken as a proportion of the fatigue limit, this gap is approximately the same for both steels.

Designing with short cracks

Given the present uncertainty with regard to the prediction of short-crack behaviour, can we evolve any guidelines which will be of use in design and

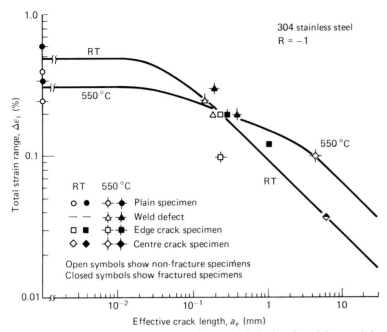

Figure 8.20 The effect of temperature on short-crack behaviour in stainless steel, from Usami[8]

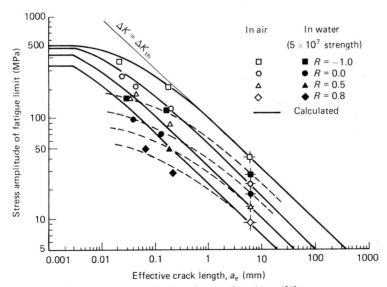

Figure 8.21 The effect of R ratio in air and water, from Usami[8]

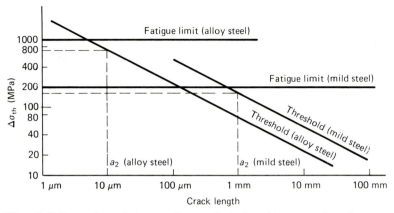

Figure 8.22 Comparison of Kitagawa diagrams for high- and low-strength steels

materials selection in situations where short cracks are important to the fatigue life of components? The following section presents some possible approaches.

Range of short-crack problems

It should be re-emphasised here that the number and type of applications in which short cracks dominate is limited. The majority of real components contain inherent defects which are larger than 1 mm, and therefore larger than a_2 for any material. Examples are weld cracks and shrinkage cavities in castings; even large inclusions will often exceed the a_2 limit. Also, even in cases where cracks may begin life below a_2, most real structures rely for their fatigue assessment on periodic NDE inspections; the size of crack which can be included in design calculations is therefore limited to the smallest detectable crack, which is rarely less than 1 mm for most NDE equipment in practical use.

However, there are a number of situations in which short-crack behaviour regularly dominates. These include:

1. Nickel-base turbine alloys in the form of single-grain blades or ultra-large grain size materials: in these very unusual alloys, cracks may be defined as short, even when equal to the component thickness.
2. High-strength materials in critical defect-free conditions: for instance, a high-strength alloy steel engine component may be manufactured to a very low inclusion content and used in a well-polished condition. In this case, minor surface defects and scratches of the order of microns in size may initiate fatigue cracking.
3. Low-strength materials in relatively critical situations such as aluminium alloy engine blocks or copper alloy heat exchangers, for which the largest defect may be submillimetre (surface roughness or small casting flaws, for example).

The above examples are all relatively small components which are intended to be used at relatively large proportions of their fatigue limits, in critical applications.

Design based on a_2 values

Faced with a situation in which the initiating defect is less than a_2 in size, the early growth rates will be difficult, if not impossible, to predict. One approach which is essentially conservative is to increase the assumed crack length to the value of a_2. This allows a safe stress to be calculated, based on normal LEFM procedures, which must be lower than the true threshold stress which will lie somewhere on the curved portion of the Kitagawa diagram.

This approach will not be overconservative, provided the crack length is not too short. Considering Figure 8.22, the largest possible error (in the case where the crack length is essentially zero, but assumed to be a_2) is about 30% for the high-strength steel and 20% for the mild steel – less than commonly used factors of safety applied to fatigue limit data. In many cases where the inherent crack sizes lie between a_1 and a_2, the degree of conservatism will be in the region of 10–15%, which is most acceptable.

The conclusion of this approach is that, provided a_2 values can be estimated accurately, short-crack behaviour can effectively be ignored in many cases without significant loss of useful strength. A slight refinement to this approach would be to assume a form for the curved portion of the Kitagawa plot. Any reasonable model, such as that of ElHaddad (see above) will generate a prediction curve which will describe real behaviour with an accuracy of better than 10% on stress level. Figure 8.23 shows

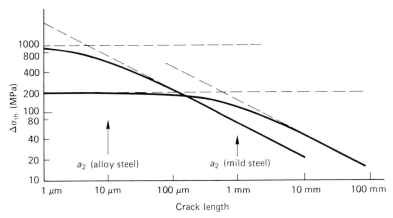

Figure 8.23 The ElHaddad prediction superimposed on Figure 8.22

curves drawn using the ElHaddad formula, equation (8.4), imposed on the plot of Figure 8.22. This curve certainly provides a conservative prediction in the region of a_2 for both of these materials; a modified form of the equation could be developed, based on known a_2 values, which would be less conservative.

The main limitation of the above approach is that it relies on accurate information of a_2 values, and therefore of short-crack data of some form. Reasonably accurate a_2 values may be estimated using either equations (8.5a) and (8.5b) above or using the stress-based method of Usami [8].

Other design approaches

Other workers[25,26] have developed probabilistic models which include a treatment of short cracks. These tend to be mathematically sophisticated but not generally linked to experimental data or crack growth mechanisms.

Kendall, James and Knott[27] have developed a conservative approach which relies on the assumption that short-crack anomalies are caused by closure only. They propose a normal LEFM approach, using the closure-free threshold value ($\Delta K_{\text{eff,th}}$ as defined in Chapter 3). This approach will accurately model the growth of short cracks which are free from closure effects and will overestimate the growth of long cracks. This approach may be very sensible in the region of a_2, where the initial, closure-free growth of the crack may dominate the life calculation. It does not, of course, model microstructural effects and therefore will not predict crack arrest or the probabilistic effects described above for shorter cracks.

References

1. Suresh, S. and Ritchie, R. O. (1984) *International Metallurgy Reviews*, **29**, 445
2. Miller, K. J. and de los Rios, E. R. (eds.) (1986) *The Behaviour of Short Fatigue Cracks*, EGF1, MEP, Bury St Edmunds, UK
3. Taylor, D. and Knott, J. F. (1982) *Fatigue of Engineering Materials and Structures*, **5**, 305
4. Blom, A. F., Hedlund, A., Zhao, W., Fathulla, A., Weiss, B. and Stickler, R. (1986) In *The Behaviour of Short Fatigue Cracks*, EGF1, MEP, Bury St Edmunds, UK, p. 37
5. Wagner, L. and Lutjering, G. (1988) In *Fatigue 87*, EMAS, Warley, UK, p. 1819
6. Lankford, J. and Hudak, S. J. (1987) *International Journal of Fatigue*, **9**, 87
7. Holm, D. K. and Blom, A. F. (1984) In *The Fourteenth Congress of the International Council of Aeronautical Sciences*, ICAS and AIAA, p. 783
8. Usami, S. (1985) In *Current Research on Fatigue Cracks*, Society of Materials Science, Japan, p. 101
9. Chan, K. S., Lankford, J. and Davidson, D. L. (1986) Transactions of the American Society of Mechanical Engineers, *Journal of Engineering Materials and Technology*, **108**, 206
10. Allen, R. J. and Sinclair, J. C. (1982) *Fatigue of Engineering Materials and Structures*, **5**, 343
11. Suresh, S. (1983) *Metallurgical Transactions*, **14A**, 2375
12. ElHaddad, M. H., Dowling, N. F., Topper, T. H. and Smith, K. N. (1980) *International Journal of Fracture*, **16**, 15
13. Taylor, D. (1986) In *The Behaviour of Short Fatigue Cracks*, EGF1, MEP, Bury St Edmunds, UK, p. 479
14. Kunio, T. and Yamada, K. (1979) ASTM STP 675, p. 342
15. de los Rios, E. R., Mohamed, H. J. and Miller, K. J. (1985) *Fatigue and Fracture of Engineering Materials and Structures*, **8**, 49
16. Taylor, D. PhD Thesis, Fatigue Crack Propagation in Nickel–Aluminium Bronze Castings, University of Cambridge p. 123
17. Pearson, S. (1975) *Engineering Fracture Mechanics*, **7**, 235
18. Miller, K. J., Mohamed, H. J. and de los Rios, E. R. (1986) In *The Behaviour of Short Fatigue Cracks*, EGF1, MEP, Bury St Edmunds, UK, p. 491
19. Lankford, J. and Davidson, D. L. (1988) In *Fatigue 87*, EMAS, Warley, UK, p. 1769
20. Cox, B. N. and Morris, W. L. (1988) In *Fatigue 87*, EMAS, Warley, UK, p. 1769
21. Chan, K. S. and Lankford, J. (1988) *Acta Metallurgica*, **36**, 193
22. Taylor, D. (1984) *Fatigue of Engineering Materials and Structures*, **7**, p. 267
23. Knott, J. F. (1988) In *Fatigue 87*, EMAS, Warley, UK, p. 497

24. Elsender, A., Gallimore, R. and Poynton, W. A. (1977) In *Fracture 77, Proceedings ICF4* (Canada, 1977) Pergamon, Oxford p. 953
25. Cox, B. N. and Morris, W. L. (1987) *Fatigue and Fracture of Engineering Materials and Structures*, **10**, 419
26. Arone, R. (1981) *Engineering Fracture Mechanics*, **14**, 189
27. Kendall, J. M., James, M. N. and Knott, J. F. (1986) In *The Behaviour of Short Fatigue Cracks*, EGF1, MEP, Bury St Edmunds, UK, p. 241

9 The effect of defect type

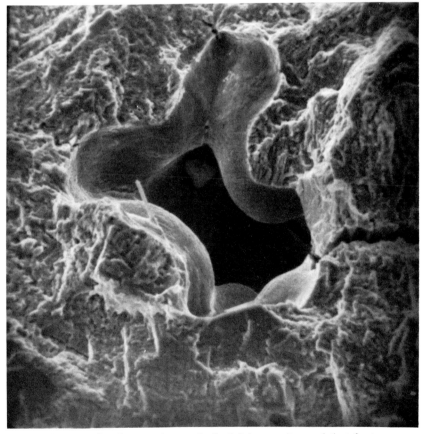

Casting defect; scanning electron micrograph of a shrinkage cavity on a fatigue fracture surface

Introduction

A major stumbling block to the use of threshold concepts in practical situations is that most cracks in engineering components are initiated at defects which only rarely approximate to the long, straight cracks used in the laboratory to define threshold behaviour.

In addition to the geometrical form of a defect there may be variations in the local stress environment (e.g. residual stress) and the local material condition (e.g. surface decarburization). This chapter considers the extent to which fracture-mechanics concepts may be used, in modified forms, to predict thresholds for cracks growing from defects.

First, the defect will be considered as a notch, exploring the problems of variation in geometrical form and stress concentration factor, K_t. Such considerations are appropriate to designed notches such as bolt holes and keyways. Subsequently, various other types of defect will be considered, including surface roughness, casting defects and inclusions. Finally, it will be shown how defect distribution and size variation may be considered by using statistical concepts.

A number of extended examples, which illustrate the problems of defect type, can be found in the case studies in Chapter 10.

Notch geometry

A considerable amount of work has been done, both experimental and theoretical, to study the development of fatigue cracks from notches. For convenience, notches may be divided into three, rather idealized, categories:

1. Very sharp notches, whose root radius, ρ, is sufficiently small that they can be considered to be equivalent to cracks of the same length: in such circumstances, at least in the ideal case, normal fracture-mechanics concepts may be applied and ΔK_{th} should be the same as that derived from crack studies.
2. Relatively blunt notches under relatively low applied stresses, generating purely elastic stress fields in the notch root region: in the case where no plastic zone is formed, and where the bluntness gives rise to a relatively large notch-tip zone in which the local stress gradient is low, it may be appropriate to assume that the stress acting on a crack at the notch root is simply equal to the far-field stress multiplied by K_t, the elastic stress concentration factor for the notch. The threshold stress range for failure then simply becomes the fatigue limit (or endurance limit) of the material divided by K_t. Blunt notches under high applied stresses are not discussed because these constitute low-cycle fatigue conditions rather than high-cycle near-threshold conditions to which this book is confined.
3. Notches which do not fit into the above categories, i.e. reasonably sharp notches subjected to sufficiently high stresses as to cause a plastic zone at the notch root: in such cases it is common to observe so-called 'non-propagating' cracks, i.e. cracks which initiate from the notch root

but arrest after growing some distance, up to a few millimetres. The important threshold is not that for initiation from the notch, but for continued growth beyond some critical length.

It is now proposed to discuss these three categories in more detail; for convenience, those notches described in section (1) above will be referred to as 'crack-like notches', those from section (2) as 'blunt notches' and those from section (3) as 'sharp notches'.

It should be borne in mind that the intention here is not to attempt a complete description of the fatigue behaviour of notches, but to define the 'threshold' conditions necessary for growth of a crack beyond the notch region. It should also be remembered that different notch geometries may give rise to the same K_t value, since K_t only describes the maximum of the elastic stress distribution. However, it appears that the general form of the near-tip elastic stress field is similar for a wide range of notch geometries[1].

Crack-like notches

It would be very useful for the design engineer to know, for a variety of materials and loading conditions, whether a given notch can be treated as a crack for the purposes of stress analysis. This one single piece of information would do a great deal to encourage the use of fracture-mechanics concepts in design. Up to now there have been only a few studies aimed at measuring, or predicting, this critical root radius, ρ_{crit}, below which a notch can be treated as a crack. The limited information available suggests that these critical values may be surprisingly large, especially in low-strength materials. For instance, Jack and Price[2], conducting S/N tests on notched specimens with varying root radii, recorded no change in behaviour for notches sharper than 250 µm in a mild steel (Figure 9.1). Similarly, Frost[3] measured critical root radii of about 600 µm in a mild steel and 150 µm in an aluminium alloy. Swanson, Thompson and Bernstein[4], in *fracture* tests on a high-strength aluminium alloy, measured a value of 20 µm for circular, V-shaped notches.

The present author, in unpublished work conducted with Knott and Turska, tested a series of notches in the quenched-and-tempered steel A533B. These notches were produced by growing fatigue cracks, blunting these cracks by application of high loads and then fully reheat-treating the specimens to remove all residual stresses from the notch roots. Figure 9.2 shows two such notches from which fatigue cracks have been initiated in subsequent testing. Two interesting results were noted: first, critical root radius values in the range 3–40 µm were measured, the critical value increasing with applied stress range. In this case ρ_{crit} was defined in terms of a notch sufficiently sharp that a crack grew from it immediately at the expected growth rate for the given ΔK value in this material.

The second observation was that, for particularly sharp notches, the initial crack growth from the notch was in fact faster than expected, and gradually decreased to the normal value. Figure 9.3 illustrates the effect of

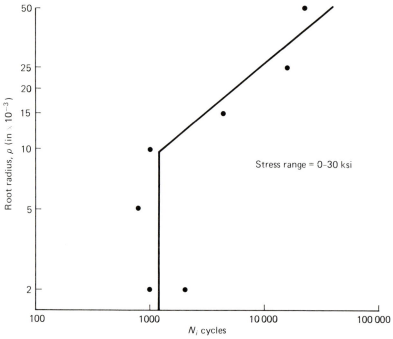

Figure 9.1 The effect of notch root radius on number of cycles to failure in a mild steel, from Jack and Price [2]

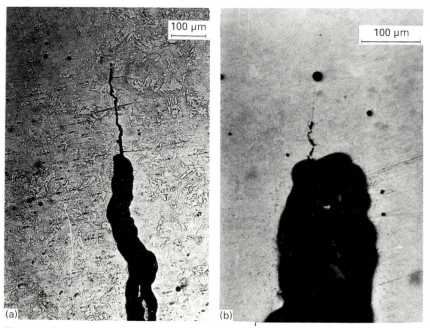

Figure 9.2 Cracks initiating from notches in A533B steel

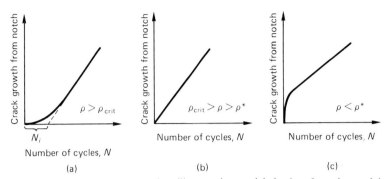

Figure 9.3 Effect of different notch radii on crack-growth behaviour from the notch in A533B steel

various root radii. This acceleration effect is probably related to closure (see Chapter 3); the notch itself will show no closure behaviour and the initial growth of the crack will be virtually closure-free until it has established a plastic wake sufficiently large to set up normal closure conditions. Thus initially, the effective ΔK value will be higher than normal, leading to faster propagation. The present tests were conducted at an R ratio of 0.4, at which a small but finite amount of closure might be expected in this steel.

The net effect of this faster growth will be to extend the crack somewhat, which might be significant if the notch length is short; in the present testing the notches were formed from long cracks and were cycled at ΔK values above the threshold. A simple theoretical prediction, which agreed well with the results, was derived by comparing the plastic zone sizes for a sharp crack and a notch. It was assumed that the critical value, ρ_{crit}, would correspond to the situation where there was a negligible difference between the crack and notch plastic zone sizes. Interestingly, this critical condition corresponded to a situation for which the plastic zone size was approximately equal to twice the notch root radius, giving the situation shown in Figure 9.4, with the plastic zone as wide as the notch itself. This is reasonable since it is expected that a crack will grow relatively easily through the plastic zone; if the plastic zone is about the same size as the root radius, a crack of plastic-zone-size length will effectively 'sharpen' the notch-tip region enough for us to consider the entire notch to be a crack.

The same theoretical approach is able to predict the 250 µm critical radius recorded by Jack and Price, the higher value in this case resulting from the lower yield strength of the material. It should be emphasized that

Figure 9.4 Relative sizes of notch and plastic zone for the condition $\rho = \rho_{crit}$

none of the above studies considered the threshold as such, all the data involving continued propagation of cracks. The above theory would predict a reduction in ρ_{crit} as the applied ΔK is reduced. The critical root radius under near-threshold conditions is expected to be in the region 1–50 μm, tending towards the smaller value for high-strength materials and to the larger value for low-strength materials.

This information will be considered in conjunction with the data described below on the effect of K_t for blunter notches which give rise to non-propagating cracks, since the sharp crack corresponds to the limit as K_t approaches infinity.

Blunt notches, low applied stress

In principle it should be relatively simple to derive a stress intensity factor for a crack situated in the purely elastic stress field of a blunt notch experiencing a far-field stress, σ, where $K_t\sigma < \sigma_y$. This analysis has been outlined by Knott[5], and becomes very simple if the crack length is very much smaller than the notch root radius, when:

$$\Delta K = K_T \sigma \sqrt{(\pi a)} \tag{9.1}$$

Even if this equation is not exact it will always provide a conservative approximation, due to the finite stress gradient in the vicinity of the notch.

It can be shown for all notch geometries that once the crack is growing, ΔK will increase with crack length despite the fact that the crack is growing into a decreasing stress field. The only remaining difficulty in this case is to define a relevant crack length for equation (9.1). Assuming that we begin with a crack-free notch, the initial crack which develops will be a 'short' crack as defined in Chapter 8. From work on plain fatigue specimens we know that, in the early stages of fatigue, cracks will initiate and grow relatively easily up to some critical length, defined in Chapter 8 as a_1, and related to both microstructural conditions (e.g. grain size) and mechanics constraints, especially the development of closure as discussed above. For low-strength materials such as mild steel, a_1 will typically have a value of 100 μm, for high-strength materials it may be less than 10 μm.

We can therefore assume that cracks up to a_1 in size already exist at the notch root. The propagation and threshold behaviour of such a crack can be treated in various ways of which the simplest is probably the use of a closure-free, 'intrinsic' threshold value, as outlined by Knott[6]. However, if the chapter on short cracks is considered it will be recognized that this corrected fracture-mechanics approach will in fact yield the same prediction for threshold *stress* range as that obtained by the much simpler procedure of dividing the fatigue limit (from an *S/N* curve) by K_t. This arises from the definition of a_1, using the Kitagawa diagram. In other words, for this type of notch it is not necessary to consider a fracture mechanics approach.

The predicted variation of threshold stress with K_t is shown in Figure 9.5, for an idealized material with a fatigue limit of 100 MPa. Two points should be noted about this prediction:

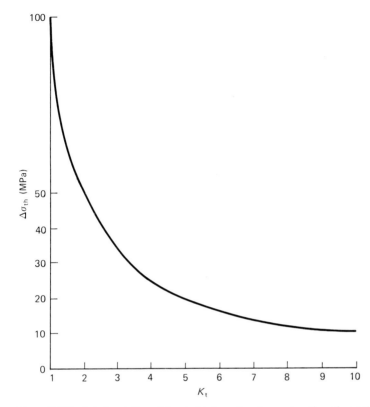

Figure 9.5 Predicted variation of threshold stress $\Delta\sigma_{th}$ for propagation of a crack from a notch – elastic solution for blunt notches.

1. Being purely elastic, it predicts a very rapid decrease in threshold stress with notch acuity. In practice, the effect may be less dramatic, though the curve is more closely approached under these near-threshold conditions than an equivalent curve in a low-cycle fatigue situation. In any case, one can rely on the fact that the curve here is a conservative prediction.
2. As will be shown below, other effects modify the curve dramatically for high K_t values, and also for the case of physically small notches.

Sharp notches; non-propagating cracks

We consider now those notches which are most common in real structures and components, i.e. relatively sharply notched design features such as thread roots and keyways and material defects such as casting porosity. The characteristic feature here is the presence of a plastic zone at the notch root during normal loading.

As stated above, it is common to observe the development of non-propagating cracks at these notches, so that our condition for a threshold from such a notch becomes a condition that a growing crack will arrest at some length rather than propagate to failure. In this sense it is accurately described as a fracture-mechanics problem.

Typical experimental results, covering the whole range of notch acuities, are shown in Figure 9.6, due to Tanaka and Akiniwa[13]. In their results they define three different types of non-propagating crack, shown on the diagram as 'through crack', and two different types of 'part-through crack'. They also record cases of total fracture and cases where no crack formed.

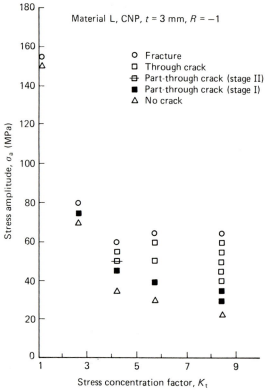

Figure 9.6 Results due to Tanaka and Akiniwa[13] on the variation of $\Delta\sigma_{th}$ with K_t

For our purposes the important feature is the stress range at which the change from non-propagating crack to fracture (i.e. continuing crack propagation) occurs. This we term the threshold stress, $\Delta\sigma_{th}$. Note that at low K_t values there is very little separation between the 'no crack' and 'fracture' data; as K_t increases there is an increasingly large region of the plot over which non-propagating cracks occur.

For very high K_t values we reach the situation defined as a 'crack-like notch', for which the analysis simplifies to a calculation of the stress *intensity* based on the notch *length* alone. For the material and notch geometry used by Tanaka and Akiniwa, this corresponds to a K_t value of about 8–9. Therefore, the 'sharp notch' region on Figure 9.5 exists in the K_t range from 3 to 9 approximately.

It is reasonable to assume that the final length of a non-propagating crack should be greater than the notch plastic zone size, since a crack which is contained within the plastic zone will experience cyclic loading at stress levels of the order of the yield strength of the material. Under such conditions a crack will always propagate. So generally the crack length at non-propagation will be somewhat larger than the plastic zone size, as shown schematically in Figure 9.7. This implies some form of elastic/plastic fracture mechanics may have to be used, based on the non-propagating crack length and the local elastic/plastic stress conditions.

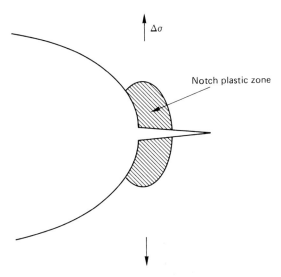

Figure 9.7 Expected position of the crack when it becomes non-propagating

Many researchers have attempted to analyse this problem (e.g. [7–17], using various corrected-LEFM approaches incorporating plasticity and closure concepts.

Prediction of threshold stress range for all notch K_t values

This section brings together the preceding sections, with the aim of developing a compete model for the variation of $\Delta\sigma_{th}$ for the whole range of notch geometries.

First it is noted that the low-K_t and very high-K_t regions can be adequately dealt with by simple analyses, as blunt, elastic notches and crack-like notches respectively. Consider the situation in which K_t is varied by varying root radius, at constant notch length, $a(n)$, which was in fact the

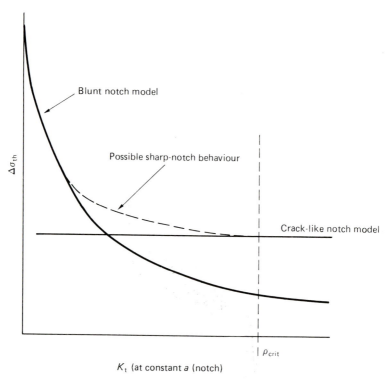

Figure 9.8 Predicted variation of $\Delta\sigma_{th}$ with K_t for constant notch length, $a(n)$

case for the data of Tanaka and Akiniwa. Figure 9.8 shows that the crack-like assumption leads to a horizontal line on the plot.

The simplest approach would be to assume that $\Delta\sigma_{th}$ takes the *highest* of the two values predicted by the two models. However, in the sharp-notch region in the centre of the plot, where neither model is valid, it is more likely that the dependence will follow something like the dotted line, with the simple models providing too low a threshold value.

However, the data of Tanaka and Akiniwa (Figure 9.5) imply that the simple two-line approach is reasonably accurate throughout. Figure 9.9 shows their data replotted with the prediction lines in place. Admittedly, the data are sparse in the low-K_t region, but the predictions are seen to give a small but acceptable underestimate of the stress for the transition from non-propagating cracks to fracture.

The reason for this may be as follows. Provided the notch root radius is not too large, the non-propagating crack will be sufficiently long to sharpen the notch-tip region, so that the notch can be considered as a crack of length equal to ($a(crack)$ + $a(notch)$), the combined notch-plus-crack length. Provided the initial notch length is great enough, the value of $a(crack)$ will be negligible. Otherwise it will tend to reduce the value of $\Delta\sigma_{th}$ somewhat. Figure 9.10 shows an idealized plot to illustrate the effect of notch length.

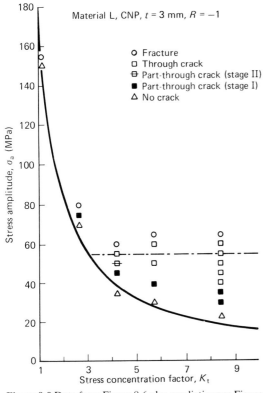

Figure 9.9 Data from Figure 9.6 plus predictions as Figure 9.8

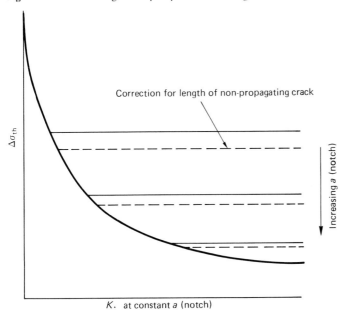

Figure 9.10 Predicted variation of $\Delta\sigma_{th}$ with K_t and with $a(n)$, showing the expected correction for the presence of a non-propagating crack at the notch root

170

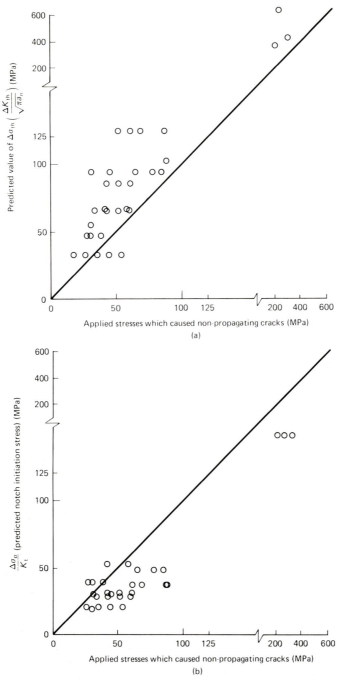

Figure 9.11 Comparison of data and predictions for growth of non-propagating cracks from notches, using data from various materials[17]. (*a*) Applied stresses known to cause non-propagating cracks compared to predicted values for $\Delta\sigma_{th}$. All data should lie above the 45° line. (*b*) Applied stresses known to cause non-propagating cracks compared to predicted values for $\Delta\sigma_0/K_t$, i.e. the threshold for initiation from the notch. Data should lie below the line

The same theoretical predictions could be represented in a number of other forms; for instance, the horizontal axis could be ρ or $a(notch)$ or some other feature of the notch geometry. Other evidence to support the validity of this approach is presented in Figure 9.11, using data from a variety of sources (in reference[17]). Figure 9.11(a) compares the predicted value of $\Delta\sigma_{th}$ with experimental applied stresses which have given rise to non-propagating cracks, for various materials. These stresses should therefore all be less than the predicted threshold value, i.e. the data points should all lie *above* the line drawn at 45°. With one exception, this is the case. It can also be seen that several materials show data which approach the predicted threshold stress, suggesting that the prediction is not overconservative. Figure 9.11(b) uses the same data, comparing values of applied stress at which non-propagating cracks were recorded, to the value of $\Delta\sigma_0/K_t$, i.e. the stress at which crack initiation from the notch is predicted. In this case the great majority of results lie below the 45° line. It seems, therefore, that crack initation and continued propagation can both be described reasonably accurately using these parameters, though of course the propagation condition is much the most important one for our purposes.

K_t values for these notches varied in the range 3–20, but in all cases the crack-like prediction was used, since it gave a higher prediction for the threshold stress than the blunt notch prediction.

Very small notches

The above simple approach is intended for use in cases of relatively macroscopic notches of simple geometry, such as bolt threads, keyways and fastener holes. Smaller stress concentrators, including casting defects (Figure 9.12) and surface roughness, are more problematic. The particular problems of rough surfaces will be considered below. In the case of small, individual defects such as casting pores, *short-crack* behaviour must be considered if the defects are smaller than the critical size a_2, defined in Chapter 8. Some experimental work has been done in this area, but the situation is complicated by the presence of residual stresses and unusual local microstructures (see below) which swamp geometrical effects. However, some workers have been able to treat defects as cracks, using the Kitigawa diagram (see Chapter 8).

Murakami and Endo[18] have studied a wide range of small notches, in the size range 5–500 μm, in various alloys. They have come to the surprisingly simple conclusion that these notches can be treated as cracks of equivalent length, and have allowed for differences in shape by the use of a very simple parameter based on the projected *area* of the notch.

Notches which are smaller in size than a_1 (defined above and in Chapter 8) should, in theory, be simple to describe. Short cracks will develop from such notches more easily than they will from other regions of the material; the problem is, then, the same as that concerning the propagation of a short crack in a plane specimen, except that the crack length is increased by the notch length. Lucas[9] presents a similar argument, coining the term 'non-damaging notches' to refer to notches smaller than a_1. He presents some evidence from an alloy steel to show that these notches do not lower

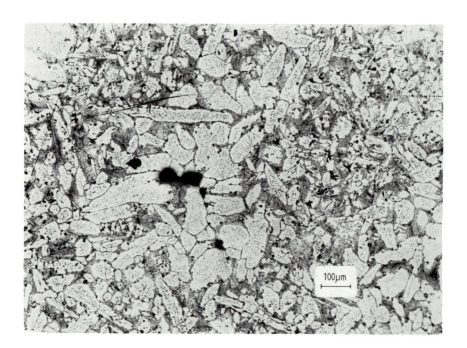

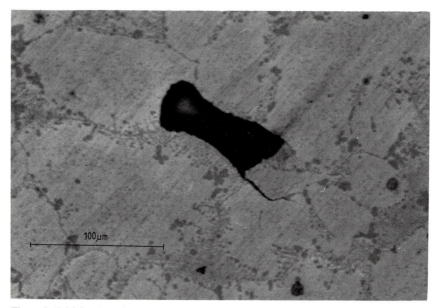

Figure 9.12 (*a*) Small casting defect with associated microstructure. (*b*) Small casting defect with fatigue crack

the threshold stress range, while larger notches with the same K_t value lower the threshold stress by up to a factor of K_t.

Effects of material properties and microstructure

Figure 9.13 illustrates the predicted variation of $\Delta\sigma_{th}$ for typical high- and low-strength steels. Since low-strength steels invariably have relatively

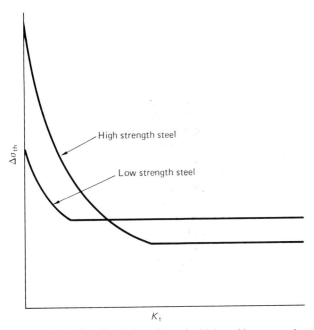

High strength steel

Low strength steel

Figure 9.13 Predicted variation of $\Delta\sigma_{th}$ for high- and low-strength steels

high ΔK_{th} values compared to high-strength steels, the lines will cross over at some K_t value. This leads to two interesting conclusions:

1. A low-strength material may be superior in situations involving relatively sharp notches. This is simply a restatement of traditional 'notch-sensitivity' concepts. In a subject such as fatigue thresholds, however, which is dominated by fracture-mechanics concepts, it is easy to forget that the low-threshold, high-strength steel *may* be the better material for a notched component if the notch acuity is not too great.
2. The low-strength steel becomes 'crack-like', i.e. it reaches the horizontal line in the diagram, at a lower K_t value than the high-strength steel. This implies that fracture-mechanics concepts may be used over a wider range of notch types for the low-strength material. This is in agreement with the higher ρ_{crit} values predicted for low-strength materials above. However, it also assumes that the simple two-line predictive model is appropriate in this case; for low-strength materials the model is liable to be more inaccurate overall than it is for strong, brittle materials.

Prediction of length of non-propagating crack

Yates and Brown[17] have attempted to predict the lengths of non-propagating cracks. Their approach assumes that the cracks are initially 'short', in the sense that the Kitagawa diagram must be used to describe their threshold characteristics (see Chapter 8), and calculates nominal K values using assumptions about the notch stress field which are essentially the same those used in the above anaylsis. It is assumed that, if the crack is small enough:

$$K = 1.12 \, K_t \sigma \sqrt{(\pi a)} \tag{9.2}$$

i.e. that the crack is loaded by the maximum notch-tip stress. As the crack grows, at some point the K value becomes described by:

$$K = \sigma \sqrt{[\pi (a + D)]} \tag{9.3}$$

i.e. the standard K equation, taking the effective crack length as the actual crack length plus the notch length, D. The complete variation of K with crack length is assumed to be that shown in Figure 9.14. By comparing this with the variation in predicted threshold with crack length, as shown on the same diagram, a kind of 'resistance curve' for fatigue crack growth is developed. It is assumed that the crack will grow until the point of intersection between the K value line and the threshold line.

A similar approach has been used by Lucas and Klesnil[15], though their formulation for the small-crack K value was slightly different. One can level certain criticisms at this approach. The method of K prediction assumes that the notch-tip stresses are elastic, which is generally not the case for those types of notches which generate non-propagating cracks, and in any case the two equations used to predict K will both be invalid in the region where they intersect – equation (9.2) – because the crack is experiencing a significant stress gradient, and equation (9.3) because the crack is too small compared to the notch size.

The model was found to predict experimental results with good accuracy, but it is significant that all the successful predictions were cases in which the intersection occurred at K values corresponding to equation (9.3), i.e. the region in which it is possible to assume that the notch can be treated as a crack.

The model ignores plasticity effects at the notch root, and therefore is forced to assume that the only reason for crack arrest is the 'short-crack' effect, causing a rise in ΔK_{th} with crack length. Despite these criticisms, however, the 'resistance curve' principle which these workers introduce is an important one, which could be extended to include an elastic/plastic analysis. However, as the above work has shown, it is not really necessary to carry out this analysis in order to predict the value of alternating stress which can be safely applied to a given notch.

The 'V' notch

The perfect V-shaped notch differs somewhat from the stress concentrators discussed above. Like the perfect crack, the perfect V notch has an infinitesimal root radius, and therefore gives rise to an elastic-stress

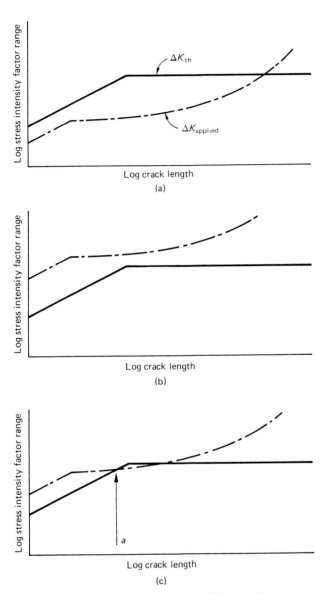

Figure 9.14 The model of Yates and Brown[17]; dotted lines represent applied ΔK, solid lines represent the variation of ΔK_{th} with crack length. The three examples are for different applied stresses; (a) no crack growth; (b) crack growth continuing to failure; (c) crack growth causing a non-propagating crack, length a

singularity. However, the stress *gradient* in the vicinity of the notch is less steep than that of a crack, and varies with notch angle. Verreman and Bailon[19] have used this approach to predict fatigue limit values for specimens containing V notches, using a modified fracture-mechanics approach which incorporates ΔK_{th} and the cyclic yield strength of the material.

Other features of defects

The preceding section treated only geometrical effects, assuming all defects to be notches of some kind. In practice, a number of other factors exert strong influences, notably residual stresses and local microstructural changes such as decarburization. Some of these problems have been outlined by Knott (9.6).

Residual stress

Residual stresses may be defined as either microscopic or macroscopic. Microscopic residual stress fields may be generated at inclusions, which are common initiators of cracks (Figure 9.15); the stress may be compressive

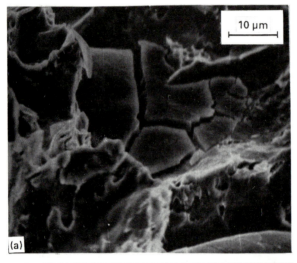

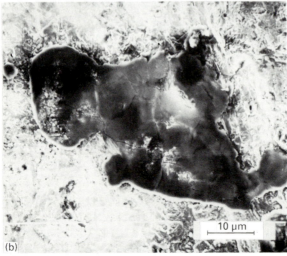

Figure 9.15 (*a*, *b*) Cracked inclusions in steels. Inclusions may be encouraged to crack by residual stress and may cause crack propagation into the surrounding material

or tensile as a result of differential contraction rates on cooling. Hydrogen cracks in non-stress-relieved welds will have monotonic plastic zones at their tips. On the macroscopic scale, most fabricating operations such as welding, forging and machining will generate some form of residual stress field. Also there are many anti-fatigue and anti-wear treatments which have the specific purpose of generating compressive residual stress fields at surfaces; these include peening, autofrettage and ballising.

The simplest approach to residual stress problems is to assume that the residual stress alters the mean stress of the imposed fatigue cycle. It is thus expected to have a significant effect unless the R ratio involved is in the closure-free region. At very high R ratios, residual stresses may be important, but can also be eliminated if plastic deformation occurs.

A number of workers have demonstrated the effects of residual stress fields on crack growth and threshold values, considering such problems as weld heat-affected zones[20] and residual stress caused by quenching, which can be either compressive or tensile in nature[21,22]. Processes such as shot-peening, nitriding and carburizing, all of which impose compressive stress, have been shown to increase thresholds and decrease growth rates[23,24].

Surface roughness

It is tempting to consider the application of the above concepts to the well-known problem of surface roughness. It has long been established that rough surfaces tend to show worse fatigue properties, but it has always proved difficult to link this effect to any defect-tolerance analysis.

The problems just mentioned, i.e. residual stress and surface microstructure, are certainly significant in this case. Shuter[25] for example, investigated surface roughness in En19 steel. He showed that 'artificial' roughness, created by forming a series of spark-erosion pits, could be analysed reasonably well as a series of crack-like notches, but surfaces produced by forging showed much worse fatigue properties for the same surface roughness owing to decarburization in the latter, which reduced the fatigue limit in proportion to the yield strength of the surface layer.

Features of rough surfaces

Obviously any surface, however well polished, contains variations in surface height if one examines it on a sufficiently fine scale (Figure 9.16). Typical surface roughness produced by metalworking operations varies from microns to hundreds of microns in scale, and therefore commonly falls within the 'short-crack' regime for cracks (see Chapter 8). However, the surface notches produced are far from crack-like; a typical surface feature is eliptical in shape with a surface-length-to-depth ratio of 10:1 (Figure 9.17), with a typical K_t value of 1.4.

An additional complication is the fact that the notches overlap, giving a series of parallel notches of randomly varying depth, sufficiently close together for their stress fields to interact.

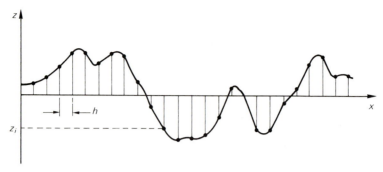

Figure 9.16 Schematic surface profile [25]

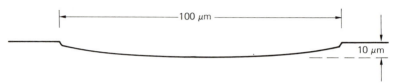

Figure 9.17 Typical geometry of a surface notch produced by a metalworking operation

It is appropriate to consider more carefully the analytical methods which can be used to describe such a surface. These methods have not been used to any great extent in relation to fatigue properties, but Moalic, Fitzpatrick and Torrance [26], for example, have used them with predictive success to deal with surface friction coefficients.

Statistical description of a surface

Digital sampling of a surface, using, for example, a Talysurf stylus machine with microprocessor data collection, will yield data in the form of a series of vertical deviations, z, from the mean height, taking h at regular sampling lengths (Figure 9.16). The commonest description of surface roughness, R_a, is given by the mean deviation which is, assuming N data points:

$$R_a = \frac{1}{N} \sum_i |z_i| \tag{9.4}$$

However, for the purpose of fatigue analysis, we are more concerned with extreme values of the distribution. Assuming that fatigue cracking commences from the deepest notch, the tendency to initiate cracking will depend on the probability of finding a certain high negative value of z in a given area of surface. Considering, then, the distribution of z values: for a typical machined surface this will be approximately Gaussian (Figure 9.18). Another commonly used roughness parameter is the root-mean-square value of z, denoted R_q:

$$R_q = \frac{1}{N} \left[\sum_i z_i^2 \right]^{1/2} \tag{9.5}$$

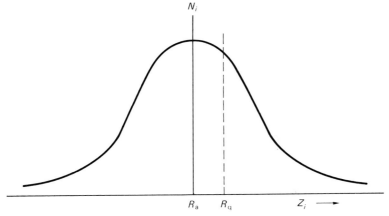

Figure 9.18 Gaussian distribution of surface height, z, showing the definition of R_a and R_q

It can also be shown that R_q corresponds to the value of z at one standard deviation from the mean of the distribution (Figure 9.18). This therefore goes some way to our requirement for an understanding of extreme values. Recent work in the author's laboratory has indicated that results from surface roughness can be plotted on a Kitagawa diagram, using R_q instead of crack length.

Ideally, one would prefer a parameter which accurately defined the 'tail' of the distribution on the negative side. However, two other statistical parameters which are easily computed digitally can give some useful information. These are the normalized averages of z^3 and z^4, denoted the 'skewness' and 'kurtosis', respectively, of the distribution.

1. *Skewness* measures the degree to which the distribution is non-symmetrical – Figure 9.19 illustrates a skewed distribution. If the skewness is such as to give a long 'tail' on the negative side, such a

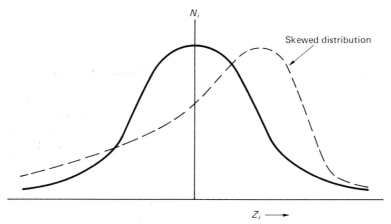

Figure 9.19 Illustration of a skewed distribution

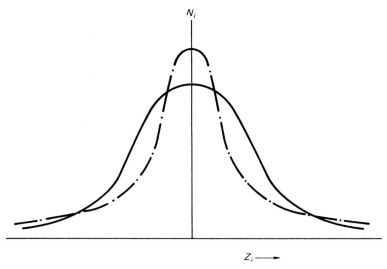

Figure 9.20 Illustration of varying kurtosis

surface has more deep notches than a Gaussian surface, and would be expected to show worse fatigue behaviour.

2. *Kurtosis* measures the 'narrowness' of the distribution; a high kurtosis implies that the distribution has relatively long 'tails', causing both deep crevices and high peaks on the surface (Figure 9.20).

The advantage of these parameters, as shown by Moalic, Fitzpatrick and Torrance[26], is that they can be computed relatively easily using a microcomputer linked to a measuring device. It is recommended that the relationship between these parameters and fatigue threshold behaviour be examined, bearing in mind the analytical difficulties outlined above, associated with blunt notches of such a small size.

Spectral analysis

Another approach, also outlined by Moalic, Fitzpatrick and Torrance[26], is to carry out a Fourier transform of the surface profile, generating a height/wavelength distribution. Since the stress-concentration at a notch depends on its width as well as its depth, relatively shallow notches of small width may be more damaging than relatively deep, wide notches. A spectral analysis such as this is also able to detect 'waviness' due to specific machining operations, and to distinguish between surfaces having the same R_a value.

It should be noted that both the above approaches are limited by the method used for profile measurement, which is usually a stylus with a finite size, and by the filtering of the data which occurs both at the level of the measuring equipment, and at the microprocessor level.

References

1. Glinka, A. and Newport, A. (1987) *International Journal of Fatigue*, **9**, 143
2. Jack, A. R. and Price, A. T. (1970) *International Journal of Fracture Mechanics*, **6**, 401
3. Frost, N. E. (1960) *Journal of Mechanical Science Engineering*, **2**, 109
4. Swanson, R. E., Thompson, A. W. and Bernstein, I. M. (1986) *Metallurgical Transactions*, **17A**, 1633
5. Knott, J. F. (1973) *Fundamentals of Fracture Mechanics*, Butterworths, London
6. Knott, J. F. (1987) In *Fatigue 87*, EMAS, Warley, UK, p. 497
7. Topper, T. H. and El Haddad, M. H. (1981) In *International Symposium on Fatigue Thresholds*, EMAS, Warley, UK, p. 7.1.
8. Hussey, I. W., Byrne, J. and Duggan, T. V. (1984) In *Fatigue 84*, EMAS, Warley, UK, p. 807
9. Lucas, P. (1987) In *Fatigue 87*, EMAS, Warley, UK, p. 719
10. Hay, E. and Brown, M. W. (1986) In *The Behaviour of Short Fatigue Cracks*, MEP, Bury St Edmunds, UK, p. 309
11. Dowling, N. E. (1979) *Fatigue of Engineering Materials and Structures*, **2**, 129
12. Hammouda, M. M. and Miller, K. J. ASTM STP 668, The American Society for Testing and Materials, p. 703
13. Tanaka, K. and Akinawa, Y. (1987) In *Fatigue 87*, EMAS, Warley, UK, p. 739
14. Tanaka, K. and Akiniwa, Y. (1988) *Engineering Fracture Mechanics*, **30**, 863
15. Lucas, P. and Klesnil, M. (1978) *Materials Science and Engineering*, **34**, 61
16. Smith, R. A. and Miller, K. J. (1978) *International Journal of Mechanical Science*, **20**, 201
17. Yates, J. R. and Brown, M. W. (1987) *Fatigue and Fracture of Engineering Materials and Structures* **10**, 187
18. Murakami, Y. and Endo, M. (1986) In *The Behaviour of Short Fatigue Cracks* EGF1, MEP, Bury St Edmunds, UK, p. 275
19. Verreman, Y. and Bailon, J-P. (1987) *Engineering Fracture Mechanics*, **28**, 773
20. James, M. N. (1987) *International Journal of Fatigue*, **9**, 179
21. Geary, W. and King, J. E. (1987) *International Journal of Fatigue*, **9**, 11
22. Plumbridge, W. J. and Knee, N. (1987) *Materials Science and Technology*, **3**, 905
23. Yan, M., Gu, M. and Liu, C. (1982) In *Fatigue Thresholds*, EMAS, Warley, UK, p. 615.
24. Clark, G. and Knott, J. F. (1977) *Metal Science*, August, p. 345
25. Shuter, D. (1986) MSc thesis, University of Dublin
26. Moalic, H., Fitzpatrick, J. A. and Torrance, A. A. (1987) *Proceedings of the Institute of Mechanical Engineers*, **201**, 321

10 Case studies and applications

Cracking in helicopter rotor lugs, Gerberich and Gunderson[9]

Introduction

If data on ΔK_{th} and near-threshold crack growth rates are valid, then the data should be usable in many practical situations to predict safe operating stresses and useful lives in service. Even in situations where LEFM parameters may be invalid, the same approach may constitute a conservative or lower-bound prediction.

Two common features of the case studies below, as indeed of most applications of threshold concepts, are the high-cycle lifetimes required and the critical nature of the components. Threshold assessment is necessary if a total life, or period between NDE inspections, in excess of a million cycles is required. This includes high-frequency loading situations, and components or structures which are relatively difficult or expensive to inspect. Likewise, the relative complexity of a threshold assessment, and the sometimes stringent recommendations which would result, such as regular maintenance periods or use of expensive alloys, restrict its use to critical components whose failure would be dangerous or expensive.

Applicability

Threshold concepts may be used with most confidence under conditions which conform to, or approximate to, the following:

(a) LEFM stress conditions;
(b) long cracks or crack-like defects;
(c) constant-amplitude loading;
(d) uniform microstructures;
(e) relatively inert environments, or well-characterized environments such as ambient air;
(f) absence of residual stress.

In practice, relatively few applications will satisfy all the above conditions, but in many cases design rules can be adopted to allow for variability and uncertainty in behaviour, in a conservative manner. The preceding chapters have considered what design rules may be appropriate to allow for different effects such as short cracks and variable-amplitude loading.

Situations which constitute problems for an LEFM-based threshold approach fall into three categories: first, there are problems associated with the defect type, its shape (especially root radius), its length (if very short), and its local environment which may contain residual stress and atypical microstructural features; second, there are difficulties associated with the assessment of corrosive environments, especially those for which there is a strong frequency effect; finally, there are problems associated with load type, of which the principal difficulties are variable-amplitude loading (especially truly random loading) and mixed-mode loading.

We will now examine some studies in which threshold concepts have been used, either to aid design and materials selection or as part of a failure analysis. Beginning with simple examples we will progress to more complex cases, finishing with reference to some more general studies which encapsulate larger applications areas and classes of materials such as welded joints and superalloys.

Critical defect size calculations

In some cases a simple calculation based on the well-known formula:

$$\Delta K = \Delta\sigma\sqrt{(\pi a)} \qquad\qquad (10.1)$$

will give an approximation, at least to an order of magnitude, of the critical crack length above which crack growth is possible for given stress conditions. This length can then be compared with the smallest detectable flaw size (using NDE methods). One example, typical of many, is provided by Gorman et al.[1], who carried out a feasibility analysis on the use of materials in a large space structure. The containing vessel of such a structure, intended for long-term human habitation in Earth orbit, is obviously critical as regards its catastrophic failure, and is relatively simple to analyse since it constitutes a pressure vessel of regular shape. Calculations for a candidate design and material showed that the largest crack in the structure which could be tolerated, based on ΔK_{th}, was 0.13 inches long. It was considered that no NDE system available could reliably detect such a small crack in such a large vessel; the material was therefore rejected.

An example of the assessment of existing known defects is provided by Pook[2], of the National Engineering Laboratories, UK. This case concerned a port block, part of the hydraulic equipment of a ship. A large block, made from thick section steel, revealed a number of planar defects after ultrasonic examination (Figure 10.1). The concern was that, in

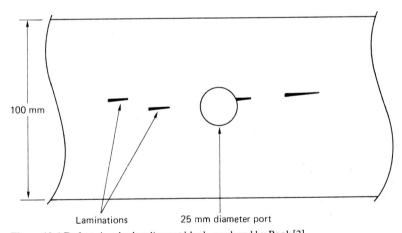

Figure 10.1 Defects in a hydraulic port block, analysed by Pook[2]

drilling ports through the block, one of these defects may be encountered, as shown in the diagram; this defect would then be under stress owing to the cyclically varying pressure through the port. A value for the threshold of the steel was measured at 7.3 MPa $\sqrt{m}$; equation (10.1) (including a small correction factor for crack geometry) calculated a value of 3.6 mm as the critical crack length. The ultrasonic method was not considered to be capable of detecting a buried defect of this kind with 100% certainty, so the block was scrapped. Since no plate material could be found with a significantly higher ΔK_{th} value the solution was to manufacture the plate via a forging route, eliminating these lamination defects.

An important factor in the assessment of NDE methods is *reliability*. It is often said that the important parameter for an NDE system is not the smallest crack which can be detected but the *largest* crack which can be *missed*. Figure 10.2 illustrates this with some data on the reliability of an

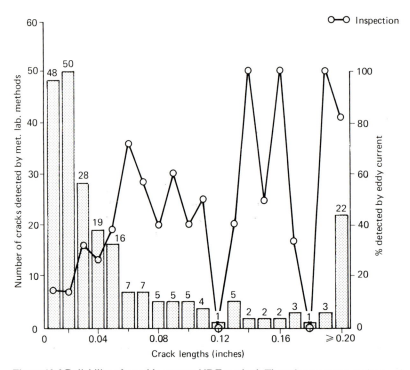

Figure 10.2 Reliability of an eddy-current NDE method. The columns measure the number of cracks actually present; the line shows the percentage success rate for detection

eddy-current method. The columns show the number of cracks which were actually present in the sample, found by sectioning and microscopy; the lines show the percentage detected by the NDE system. Even quite large cracks can only be detected with a certainty of about 80%, which would be unsatisfactory for any critical component.

Bone cement failure: testing a hypothesis

This example, taken from work in the author's laboratories, involves the materials used in the artifical hip joint. The hip joint prosthesis is now in common use for restoring mobility to patients suffering from arthritis and other diseases. As Figures 10.3 and 10.4 show, the prosthesis consists of a metal ball and stem, the stem being located down the centre of the femur

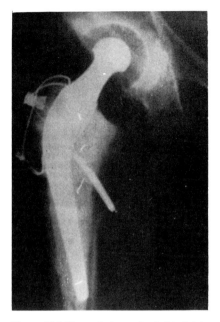

Figure 10.3 X-ray photograph of a hip joint prosthesis in place

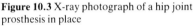

Figure 10.4 Schematic illustration of the artificial hip joint

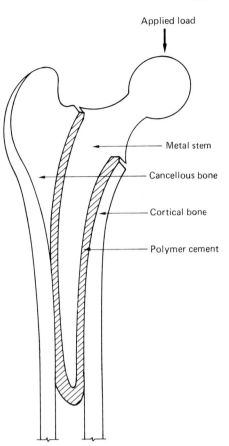

(thigh bone) and the ball locating, usually through a polymeric 'cup', into the socket of the pelvis.

In the femur, the metal stem is held in place using a bone cement. This cement, which is PMMA, anchors the metal in place and fills up any irregularities in the surface of the bone, but has little adhesive function. After a number of years of use, these prostheses frequently show signs of loosening, causing movement of the metal stem within the femur, which can eventually lead to stem fracture by fatigue. More often the loosening itself causes pain and revision surgery is required.

The hypothesis which we were required to address was that this loosening was being caused by fatigue cracking in the cement material. If it could be shown that this hypothesis was feasible, then a programme of research would be undertaken to improve the fatigue properties of the cement; otherwise, one would look instead to other causes of failure. At this stage, then, it was not necessary to prove that fatigue was the cause of the trouble, but to indicate quantitatively whether fatigue was one possible cause. Limitations of time and facilities dictated an essentially 'pencil-and-paper' approach, with assistance from existing finite element software and a few short-term mechanical tests.

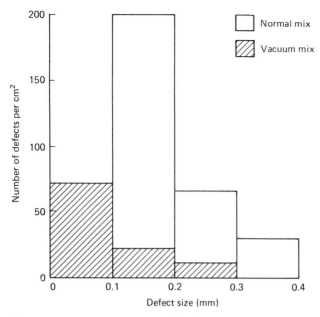

Figure 10.5 Defect population of PMMA bone cement

The cement as used contains many defects, including air bubbles and delaminations; Figure 10.5 shows the results of a defect population count. Fracture tests showed that brittle fracture could proceed from these inherent defects and that an LEFM procedure could be used, with some modifications to allow for defect shape[3], giving a consistent K_{Ic} value of 1.6 MPa $\sqrt{m}$.

The results of a three-dimensional finite-element stress analysis are shown in Figure 10.6, for normal loading through the femoral head. Figure 10.6(a) plots principal stress contours in the metal stem (a Cr–Co alloy); Figure 10(b) shows the same for the cement tube; note the two high-stress regions of 2.9–3.5 MPa and 4.6–5.5 MPa.

Assuming failure by propagation of an existing defect, the maximum tensile stress in Figure 10.6(b) can be combined with the maximum possible defect size, which is assumed to be 3 mm, the average thickness of the cement tube. Taking a value of 4 MPa (which is probably not the highest value) gives an applied stress intensity of about 0.4 MPa $\sqrt{m}$. No threshold data for this material was available; Figure 10.7 shows available results for bone cement at higher ΔK values[4] plus one of the few available threshold curves for a polymer, this being Usami's data on epoxy[5]. From this data it seems likely that 0.4 MPa $\sqrt{m}$ will be above the threshold for the bone cement.

Thus the conclusions of this feasibility study were:

1. Fatigue failure by propagation of existing defects is possible for the maximum size of defect known to be present.
2. Other polymers, such as epoxy, may be defect tolerant for the same defect size.

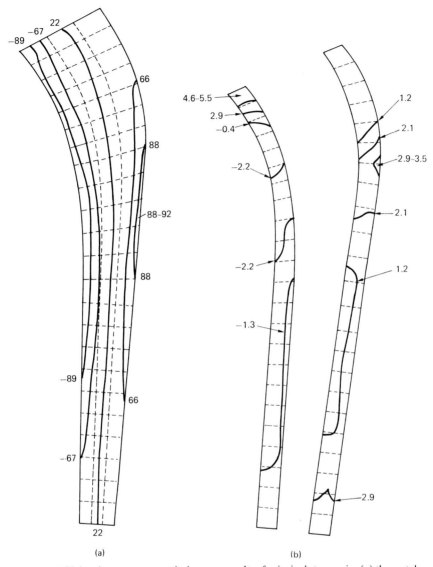

Figure 10.6 Finite element stress analysis; some results of principal stresses in: (*a*) the metal stem; (*b*) the cement mantle. Stresses in MPa; negative values indicate compression. Courtesy of P.Prendergast

Development work was therefore concentrated on improving the fatigue crack propagation behaviour of the cement, either by using a different polymer, by reducing the defect population, or by reinforcing the PMMA cement with fibres. Eventually the reinforcing route was successfully adopted.

The significance of this example is that a relatively short study was able to establish the *possible* importance of fatigue failure, indicating a line of action for a longer study.

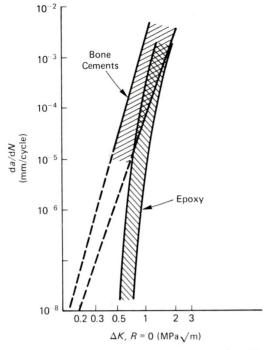

Figure 10.7 Crack propagation data on epoxy resin and bone cements

Heart valve material: assessment of short cracks

Another example from the area of 'biomaterials' is the work of Ritchie and Lubock[6], assessing a cobalt alloy for use in an artificial heart valve. Two small metal struts of complex shape are required in this design to support the operation of a valve disc (Figure 10.8) which pivots to open the valve. A thorough testing programme established both S/N data (Figure 10.9(a)) and da/dN data (Figure 10.9(b)) in a representative environment. Stress analysis comprised pulse duplicator studies to establish maximum applied loads, and two-dimensional finite-element analysis; stresses were predicted to vary between 0 and 76 MPa. Detailed stress intensity calculations established K values for a variety of possible crack geometries.

A critical defect size of 800 µm was calculated to correspond to a stress intensity of ΔK_{th}; applying a factor of safety the researchers concluded that NDE capable of detecting 500 µm flaws was required. This is a stringent requirement but not impossible given the small size of the components and the fact that the flaws would be surface-breaking.

However, given the small size of cracks involved, it was necessary to check whether cracks of this size conformed to normal LEFM behaviour, or whether they would behave anomalously as 'short cracks', as described

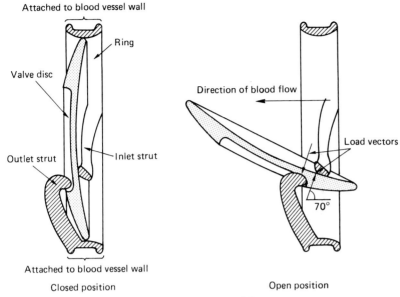

Figure 10.8 Components of the heart valve prosthesis[6]

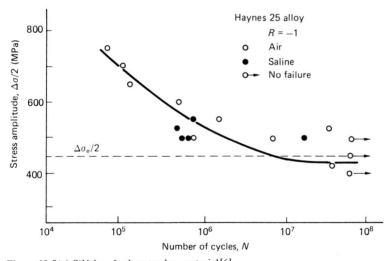

Figure 10.9(a) S/N data for heart valve material[6]

in Chapter 8. Experimental testing of short cracks was not carried out, but using the threshold and fatigue limit values, a Kitagawa diagram was constructed (Figure 10.10). The value for a_0, taken as the intersection of the two lines on the diagram, is 75 μm; this gives confidence that cracks of 500 μm or more in length will behave as normal, long cracks.

It is important to make some estimate of the short-crack parameters for any situation involving submillimetre cracks since, as Chapter 8 shows, some materials will show short-crack anomalies up to 1 mm.

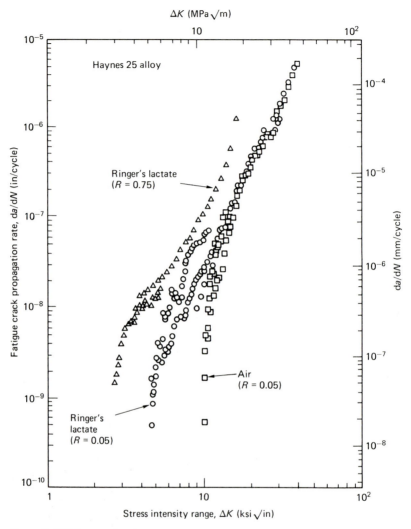

Figure 10.9(b) Propagation data for heart valve material[6]

Ship propellers: failure from casting defects

A problem in which the author was involved a few years ago concerned fatigue failure in ships' propeller blades[7] made from an aluminium bronze alloy. The propellers of ships such as oil tankers can be very large (Figure 10.11) and are generally made as single castings, from copper-based alloys which are chosen for their resistance to stress corrosion cracking and cavitation. A propeller rotating in still water will, theoretically, experience a constant force, but the presence of the ship's hull shields the propeller for part of its rotation, a higher force being imposed on each blade at the bottom of its circle of rotation. This gives rise

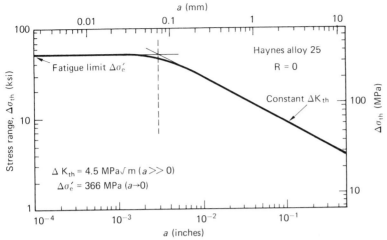

Figure 10.10 Kitagawa diagram for heart valve material [6]

Figure 10.11 The propeller of a large ship

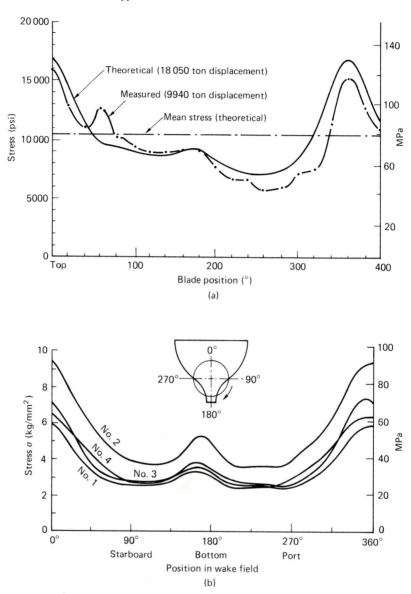

Figure 10.12 Typical service stresses on propeller blades, as a function of rotational position

to cyclic stresses of the same order of frequency as the rotation frequency (about 1 Hz), the pattern of which varies from ship to ship (Figure 10.12), but may be approximated by a sine wave with an R ratio of 0.5.

Threshold testing

The alloy used displayed normal threshold behaviour when tested in air at relatively high frequencies (Figure 10.13). However, it was discovered that

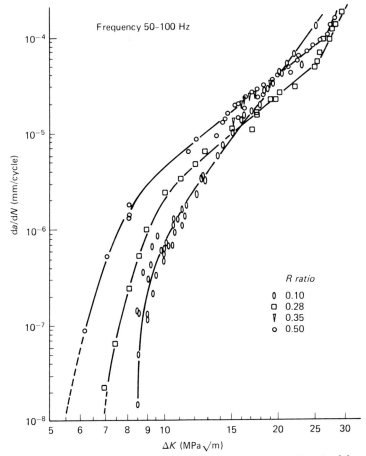

Figure 10.13 Crack propagation data for aluminium bronze propeller alloy[7]

da/dN increased as the frequency was reduced in the range 100Hz–1Hz (Figure 10.14). Initially, this was thought to be caused by a corrosion-fatigue effect in air, but it was eventually shown to be linked to a strain-rate sensitivity in the material[8], explainable by dislocation dynamics. A further corrosion-fatigue effect was identified in simulated seawater testing (Figure 10.15), which also was more significant at lower frequencies.

Since it was not feasible to establish threshold values for the material at 1 Hz in seawater, some estimate had to be made to allow for the decrease in threshold caused by these two effects. The dislocation dynamics analysis gave confidence for a prediction that the 1 Hz, air, line would parallel the 100 Hz, air, line down to ΔK_{th} (Figure 10.16). The corrosion fatigue effect was more difficult to predict, and still remains so today. It is reasonable to assume behaviour which lies between the two prediction lines in Figure 10.16, though this gives a wide margin of error.

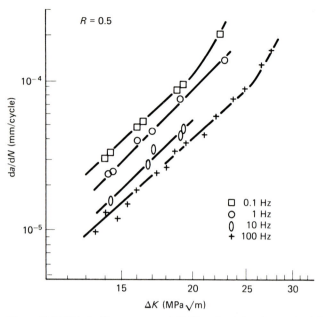

Figure 10.14 Effect of frequency on da/dN for aluminium bronze in air [8]

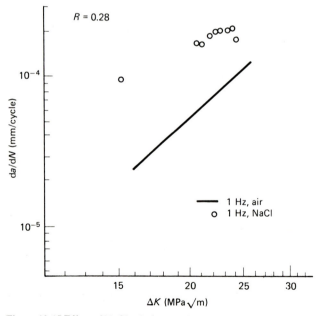

Figure 10.15 Effect of NaCl solution at 1 Hz, aluminium bronze [7]

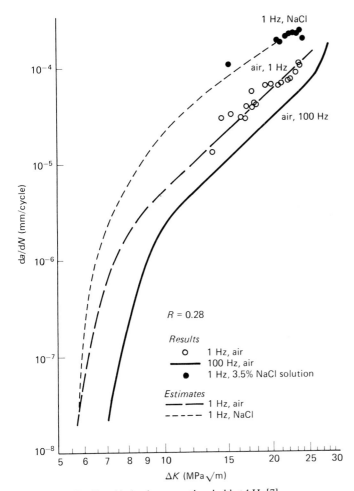

Figure 10.16 Predicted behaviour near threshold at 1 Hz [7]

Defect analysis

Material was examined which had been produced by a simulated casting representative of the large sand castings used for propellers. This showed a large number of small casting defects, mostly shrinkage cavities, with a population distribution as shown in Figure 10.17. These cavities showed sharp, re-entrant shapes (Figure 10.18); tests conducted on unnotched specimens showed that cracks initiated easily from the corners of these defects (Figure 10.19). Growth-rate measurements on such cracks showed that:

1. No initiation period was required, crack growth proceeding from the first stress cycle.
2. The defects could be treated as cracks in a fracture mechanics analysis.

198

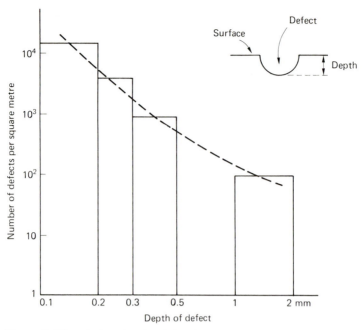

Figure 10.17 Population of casting defects in aluminium bronze, representative of a propeller casting

Figure 10.18 Typical shape of a shrinkage cavity

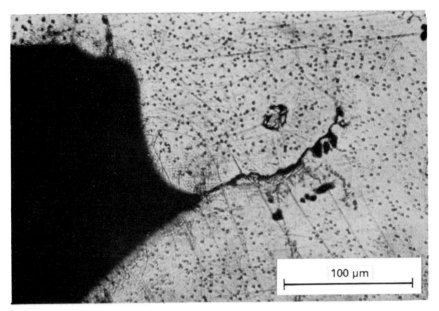

Figure 10.19 Fatigue crack initiating from a sharp corner of a shrinkage cavity

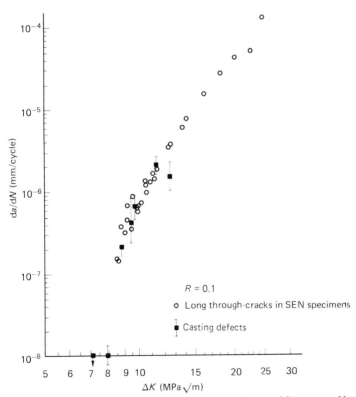

Figure 10.20 Crack propagation data from casting defects and from normal long cracks [7]

This second point is illustrated in Figure 10.20, which compares crack growth data for conventional long cracks with results from casting defects, calculating ΔK assuming that the defect size was equal to a crack length. Clearly, these defects have sufficiently sharp root radii to qualify as cracks, according to the analyses outlined in Chapter 9.

Short-crack assessment

It was not possible to estimate short-crack behaviour through a Kitagawa diagram as reported in the work of Ritchie and Lubock above, because all the material available contained casting defects. It was therefore not possible to measure the fatigue limit for defect-free material. Therefore, a large measurement programme was initiated, collecting data on cracks growing from casting defects of all sizes. Some of this data has been reported elsewhere in this book (e.g. Chapter 8); in summary it was found

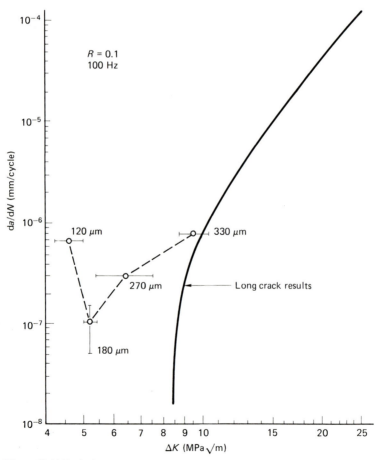

Figure 10.21 Typical propagation behaviour of a crack growing from a small defect, compared to long-crack behaviour [7]

that the critical crack length, a_2 was 300 μm. The typical behaviour of a smaller defect is shown in Figure 10.21. In this material the short-crack behaviour clearly had a microstructural origin, with grain boundaries and other features affecting growth rates of the short cracks.

Life calculations

Normal integration methods were used to estimate the number of cycles to propagate a crack to failure for a given applied stress and initial defect size, leading to a predicted S/N curve for which the fatigue limit corresponded to the threshold. The short-crack limit of 300 μm was smaller than the expected defect sizes in practice; therefore short-crack anomalies were not involved in the calculations. Initially, to test the method and data, predictions were made of some S/N tests which had been conducted by other workers using the same alloy, testing in air. As Figure 10.22 shows,

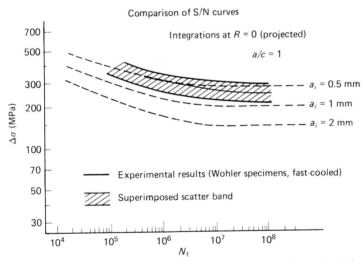

Figure 10.22 Comparison of S/N data and predictions, using various initial defect sizes

good predictions were achieved, assuming initial defect sizes in the range 0.5–1 mm, which was reasonable for the large, cast specimens used. The scatter in experimental data could also be explained in terms of variation in size of the initiating defect.

The same approach was used to predict fatigue behaviour of actual propellers. Figure 10.23 compares predicted S/N curves with known stress levels in service for various propellers. Considerable uncertainty is introduced by the need to allow for the effects of frequency and corrosion-fatigue discussed above, but it seems that the threshold condition for crack propagation is probably being exceeded for the more highly stressed propellers.

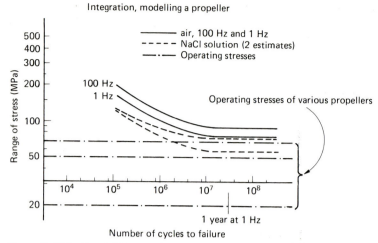

Figure 10.23 Predicted *S/N* curves for ship propellers made from this aluminium bronze alloy. The horizontal lines show actual service stresses for various propellers

Conclusion of the investigation

What has the above, rather exhaustive, investigation, been able to achieve? First, it has demonstrated that fatigue failure is a possible mechanism for this component in normal service. This overrules the argument that failures only occurred owing to misuse of the component by, for example, poor-quality weld repair or damage during installation.

Second, the analysis points to casting defects as the culprits. It is difficult at first to appreciate that a 10-foot long propeller blade may fail owing to the presence of a casting defect a few millimetres in size. A simple calculation of the effect of defect size shows, as in the *S/N* plots in Figure 10.24, that modest reductions in defect size can result in huge benefits in

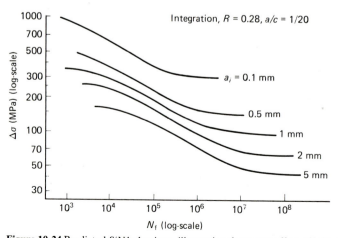

Figure 10.24 Predicted *S/N* behaviour, illustrating the strong effect of initial defect size on life

terms of increased life. In such a large structure, made by sand casting, it is impossible to eliminate defects altogether, but a careful inspection and repair procedure, concentrated on the critical high-stress locations at the blade roots, would be a practical alternative.

Helicopter rotor lugs

A final example of this type of problem is provided by Gerberich and Gunderson[9] who reanalysed the problem of cracking in the lugs of helicopter rotor blades (Figure 10.25), using data from Salkind and Lucas[10] and others.

Figure 10.25 Cracking in helicoptor rotor blade lugs[9]

The problem is similar to the previous one, except that in this case cracks are known to arise by fretting fatigue. The K calculation is somewhat more complex also; the full analytical solution, due to Newman[11], is:

$$\Delta K = \Delta\sigma \sqrt{(\pi a)} f_w f_b \left(\sec \frac{\pi D}{2W} \right)^{1/2} G$$

where

$$f_w = \left[\sec \left(\frac{\pi}{2} \times \frac{D + a}{w - a} \right) \right]^{\frac{1}{2}}$$

$$f_b = 0.707 - 0.18\lambda + 6.55\lambda^2 - 10.54\lambda^3 + 6.48\lambda^4 \qquad (10.2)$$

$$\lambda = \frac{1}{1 + (2a/r)}$$

$$G = 0.5 + \frac{w}{\pi(D + a)} \left[\frac{D}{D + 2a} \right]^{\frac{1}{2}}$$

using the lug geometry shown in Figure 10.26. However, it can be shown that, for the lug configuration of interest, this can be approximated by the simpler expression:

$$\Delta K = \Delta\sigma\sqrt{(\pi a)}\, Y$$

where

$$Y = \frac{8.159}{[1 + 20(a/r)]^{\frac{1}{4}}} \qquad (10.3)$$

provided a/r is kept relatively small.

Using known data for the Ti–6Al–4V alloy, Gerberich and Gunderson computed an S/N curve and compared this with known data (Figure 10.27). In their case the fit is reasonably good at high stresses but unduly conservative at low stresses. If their data are re-examined, however, it is

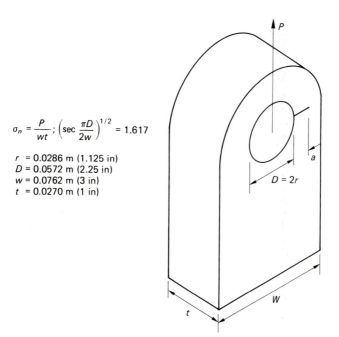

$$\sigma_n = \frac{P}{wt}; \left(\sec \frac{\pi D}{2w} \right)^{1/2} = 1.617$$

r = 0.0286 m (1.125 in)
D = 0.0572 m (2.25 in)
w = 0.0762 m (3 in)
t = 0.0270 m (1 in)

Figure 10.26 Lug geometry, for equation (10.2)

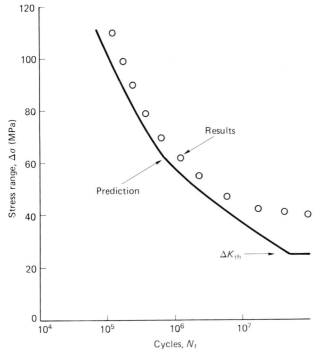

Figure 10.27 Comparison of data and predictions for lug failures[9]

found that the experimental points reach a fatigue limit at a stress of 40 MPa, which corresponds to a threshold of about 3 MPa√m, which is a typical figure for ΔK_{th} values of this alloy at this R ratio (0.43). It is possible, therefore, that they were using too low a value for ΔK_{th}.

Another shortcoming of the approach is that it does not allow for short-crack behaviour. The assumed initial crack length (produced by fretting fatigue) is 0.03 mm, this being chosen because it corresponds to the grain size of the material. The implication (see Chapter 8) is that short-crack behaviour must occur for this crack in the early stages of its growth, unless the material is in a microstructural condition which is dominated by precipitates or other smaller features. In fact, Brown and Taylor[12], testing commercial heat treatments of this alloy, showed that short-crack behaviour persists to surprisingly large crack lengths, up to 0.5 mm in material with a grain size smaller than 0.03 mm. The effect of this short-crack behaviour will be to lower the effective threshold and thus lower the predicted fatigue limit considerably. In practice, it is likely that larger cracks will develop by the fretting mechanism; microscopic examination of failed lugs will probably establish the sizes of these cracks, since they are likely to be strongly marked by corrosion products. In this case the practical solution found was to eliminate fretting fatigue by the use of sacrificial liners in the lugs.

Other application areas

The following section will treat a number of examples which are not specific cases, but general applications areas for which the use of threshold concepts presents a challenge and a possible solution to lifing problems.

Welds

Welded joints are possibly the most important application for fatigue crack growth analysis, first because a large proportion of service failures occur from welded joints, and second because weld defects from which cracking begins often take the form of cracks (e.g. heat-affected zone stress-relief cracks) or crack-like notches (e.g. undercuts, inclusions).

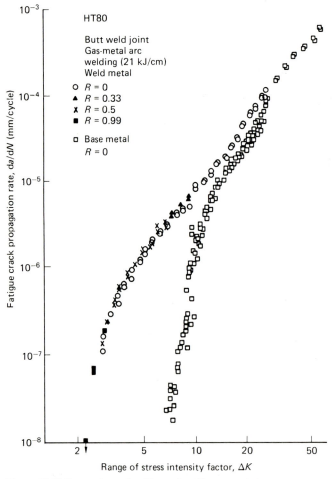

Figure 10.28 Comparison of weld metal and base metal, from a butt weld, Ohta *et al.* [13]

As yet, however, it has been difficult to apply threshold concepts with any confidence in this area, for a number of reasons: first because of the large and complex role of residual stress; second because of the varying properties of the weld metal and heat-affected zone through which the crack may be growing; and finally because of the complex geometry (and therefore the complex stress field) of most real joints.

Despite these problems, some important observations have been made. For instance, Ohta *et al.* [13] and the National Research Institute for Metals, Japan[14] have provided convincing evidence for a simple approach to butt welds in steels. They showed that cracks growing parallel to the weld seam gave much lower thresholds than base metal (Figure 10.28). More significantly, they found very similar growth rates and thresholds for a wide range of R ratios in both weld metal and heat-affected

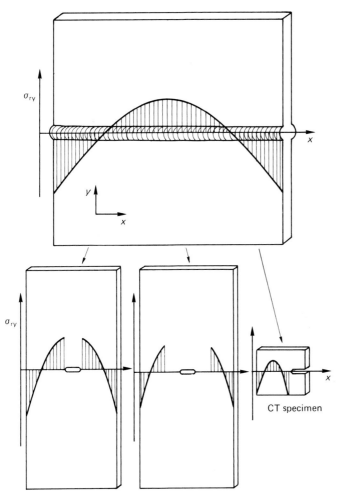

Figure 10.29 Pattern of residual stresses in a butt weld, showing the effect of cutting a CT specimen[13]

zones (HAZ). This would appear to be due to tensile residual stress at the crack tip, the effect of which is to prevent closure, tending to reduce all results to a 'closure-free' or 'intrinsic' threshold with a value of $2-3\,MPa\sqrt{m}$. Figure 10.29 shows schematically the expected pattern of residual stress; Ohta *et al.* make the important observation that for their tests, conducted on centre-cracked plates, the crack tip is always subject to residual tensile stress, but for a CT specimen machined from the same plate, the readjustment of residual stress would always place the crack tip in compression, giving uncharacteristically good results (see Figure 10.29). This work suggests that, at least for this simple weld geometry, a practical, conservative design approach may be possible.

Considering more complex joint geometries, Usami[15] has developed the use of short-crack and threshold concepts to deal with weld-toe defects, using a notch analysis which is a modified form of conventional short-crack approaches.

Turbine materials

Some unique problems are presented by materials used in aircraft engines and electricity-generating turbines. Of special concern are the nickel-base superalloys, though titanium and ferrous alloys are also used. In jet engines, the continuing demands for increased thrust and improved power/weight ratios involve materials being used in situations which combine temperatures close to the melting point with stresses close to the yield strength. Some of these materials are produced in the form of single crystals or possessing very large grain sizes for the purpose of creep resistance, introducing the anomalies associated with 'short cracks' (Chapter 8) for cracks which may be physically large compared to component dimensions.

Electricity-generating turbines suffer the additional problems of corrosion, e.g. in high-pressure steam. In both applications inspection is costly, giving a strong economic drive to increase operating periods between inspections.

The effects of microstructure and other material variables for these alloys are covered in Chapter 5. Mention here will be made of the work of Knott, King and co-workers[16–18] who have analysed nickel-base alloys from a threshold point of view.

Aeroengine turbines

A typical high-pressure stage consists of a disc of outer diameter 0.75 m, bore 0.2 m and thickness 0.1 m, to which is attached a number of blades, usually with fir-tree roots. The most significant fatigue cycles are expected to be the infrequent, start/stop cycles during which the engine temperature varies from ambient to 600°C; stresses arise from centrifugal forces and from thermal expansion.

Lifing procedures

Conventional design is based on full-scale spin-rig tests in which 500 discs may be failed, generating a statistical distribution of failure times. Typically, the safe life is reckoned to be the mean minus three standard deviations, and the US Federal Aviation Authority requires that the usable life should be one-third of this safe life.

A fracture-mechanics approach should be able to remove some of the uncertainty and conservatism of the above method, and may offer the possibility of a 'retirement for cause' approach, in which components are not removed until damage in the form of cracks is detected, rather than a

Figure 10.30 (a, b) Scanning electron micrographs of a fatigue fracture surface from a failed compressor blade root. At the origin of failure is a small inclusion, shown magnified in Figure 10.30(b)

system in which components are automatically removed once their design lives are exceeded. It is known that failures generally occur from inclusions in the nickel and titanium alloys, which are made to extremely high levels of cleanliness. Figure 10.30 shows a fracture surface from a Ti–Al–V compressor blade in which a catastrophic fatigue failure was initiated by the presence of an inclusion which had a length of only 25 μm.

Knott[17] estimated initial defect sizes of 0.18 mm and 50 μm for a life of 10 000 cycles in the alloys astroloy and rene 95, respectively, assuming cyclic loading to the proof stress but using an LEFM calculation. The use of a powder route for manufacture of these alloys allows the possibility of very stringent controls on inclusion size, through the use of sieving procedures for the initial powder. This means that, though the above crack sizes are impossible to detect using NDE methods, confidence can be ensured through control of the metallurgical process. In practice it is possible to guarantee a maximum inclusion size of 40 μm, but the high cost and low yield of the process become significant factors. This so-called 'process control' places the emphasis for component integrity on the skill of the metallurgist, rather than the NDE engineer.

Knott's analysis tends to be pessimistic when compared to the results of spin-rig tests; one reason may be the presence of beneficial residual stresses around the inclusions, caused by differential contraction during cooling. Tsubota, King and Knott[16] showed that this residual stress gave rise to a stress intensity of:

$$K_{res} = 0.058 P (2\pi r_i)^{1/2} \qquad (10.4)$$

where r_i is the radius of the inclusion and P is the residual pressure on the inclusion, which will vary with temperature.

Statistical treatments

The introduction of probabilistic concepts into fatigue predictions is well known as a method of treating S/N data. The scatter in number of cycles to failure at a given applied $\Delta\sigma$ can be represented by a probabilistic distribution such as Gaussian (Figure 10.31), allowing percentage lines to be placed within the scatter band which define the probability of failure in a given time. An alternative approach is to plot probability of failure as a function of stress, for a given number of cycles (Figure 10.32). These probability figures may provide input into design calculations. For example, highly critical, structural components may be specified to a very low failure probability, but the use of two such components in a fail-safe arrangement will reduce the specification to the probability of both failing within a given period. At the opposite end of the scale, mass-produced, non-critical components may be specified with a relatively high probability of failure, balancing manufacturing costs against the costs of failure and replacement.

Knott[19] has applied probability concepts to fatigue crack growth from defects, assuming a Weibull distribution of initial defect sizes and using a threshold value to predict an endurance limit. A major difficulty of such approaches is the specification of the initial defect distribution. Once this

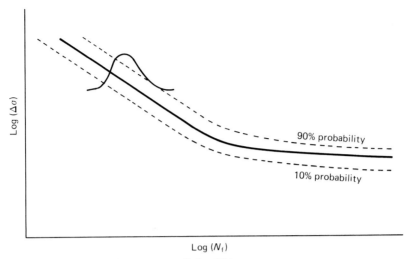

Figure 10.31 Probabilistic concepts applied to *S/N* curves

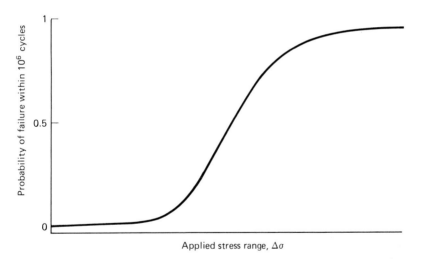

Figure 10.32 Probability of failure in a given number of cycles, as a function of stress level

has been done, however, it is possible to incorporate effects such as specimen size and loading type through the concept of a stressed volume. In a bend specimen, for instance, tensile stress falls off as one moves away from the surface, but the probability of encountering a large defect increases as more of the specimen volume is sampled.

In Chapter 8 it was shown how the behaviour of short cracks could also be described in probabilistic terms, introducing the concept that a crack of a given length, subjected to a given applied stress has a certain probability of growth, which may initially decrease with increasing crack length. The presence of a large number of short cracks, which is generally the case for

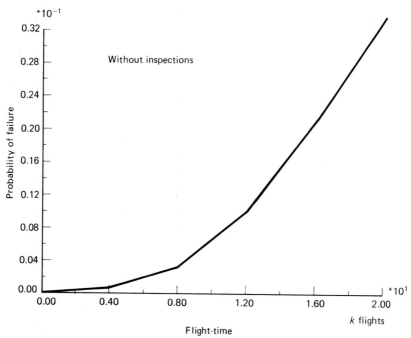

Figure 10.33 Predicted failure probability, as a function of number of flights, for an aircraft component, from Palmberg, Blom and Eggwertz[20]

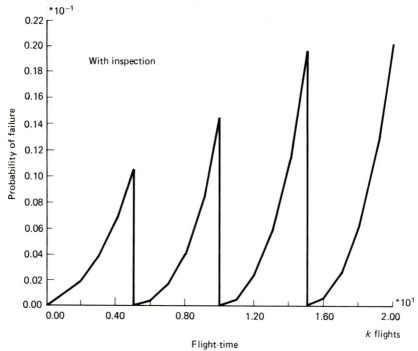

Figure 10.34 Predicted failure probability as in Figure 10.33, introducing regular inspection periods[20]

low-strength materials and materials tested close to their fatigue limits, makes a probabilistic analysis very attractive.

Palmberg, Blom and Eggwertz[20] have surveyed the use of probabilistic concepts for the description of fatigue reliability, and have developed a lifing procedure for aircraft components in which probability functions are assigned not only to the initial defect size and the stress pattern, but also to the reliability of NDE procedures. The model does not include near-threshold conditions, being concerned with lives of the order of 100 000 flights, but the use of crack propagation concepts is well illustrated. Figures 10.33 and 10.34, taken from their model predictions, show the effect of regular inspection periods. Note the difference in scale on the probability axis; the probability of failure rises successively despite the constant frequency of inspection, owing to the limited reliability of the NDE method and the increased probability of high load excursions.

Rough surfaces

The prediction of fatigue behaviour of rough surfaces presents a challenge which may be addressed using some of the above concepts. It is well known that rough surfaces tend to have worse properties than smooth ones, and that manufacturing operations produce surfaces which vary greatly in their fatigue performance. Three features of a surface may be significant:

(a) surface roughness;
(b) surface residual stresses;
(c) surface microstructure (e.g. decarburization).

Residual stresses may be deliberately introduced to improve fatigue properties (e.g. by shot peening) but most common machining operations give rise to significant residual stresses.

Optimization is usually possible by careful polishing and stress relief, but this will not be economic for many components. An accurate fatigue analysis for a given surface in the as-machined, as-forged or as-cast state will enable the designer to decide whether a surface-improving operation has to be carried out.

In principle, the high-cycle fatigue behaviour of such surfaces could be analysed using threshold concepts. In practice, the analysis involves a number of problem areas which have been discussed above and in earlier chapters.

Probabilistic concepts are clearly needed, since failure will be assumed to occur from the worst available defect, i.e. the deepest surface pit. This pit will generally be notch-like, in the sense of having a high ratio of root radius to depth. The pit may also qualify as a short crack; typical machining operations produce pits of the order of $10-100 \mu m$, which will be short (as defined in Chapter 8) for all but the highest-strength materials.

However, recent work in the author's laboratory, described in Chapter 9, suggests that an approach combining short cracks and statistical concepts may be capable of giving useful predictions.

References

1. Gorman, J., Stagliano, T. R., Orringer, O. and McCarthy, J. F. (1977) In *Case Studies in Fracture Mechanics* (eds. Rich and Cartwright), US Army Mechanics Res. Laboratory
2. Pook, L. P. (1977) In *Case Studies in Fracture Mechanics* (eds. Rich and Cartwright), US Army Mechanics Res. Laboratory, p. 4
3. Taylor, D. (1988) In *Fatigue '87*, EMAS, Warley, UK, p. 1353
4. Rimnac, C. M., Wright, T. M. and McGill, D. L. (1986) *Journal of Bone Joint Surgery*, **68A**, p. 281
5. Usami, S. (1982) In *Fatigue Thresholds*, EMAS, Warley, UK, p. 205
6. Ritchie, R. O. and Lubock, P. (1986) *Journal of Biomechanical Engineering* **108**, p. 153
7. Taylor D. and Knott, J. F. (1982) *Metals Technology*, **9**, 221
8. Taylor, D. and Knott, J. F. (1984) In *The Sixth International Conference on Fracture*, Pergamon, Oxford, p. 1759
9. Gerberich, W. W. and Gunderson, A. W. (1982). In *Application of Fracture Mechanics for Selection of Metallic Structural Materials*, ASM, USA, p. 350
10. Salkind, M. J. and Lucas, J. J. (1972). In *Corrosion Fatigue* NACE-2, The National Association of Corrosion Engineers, Houston, p. 627
11. Newman, J. C. (1976) NASA TN D-8244
12. Brown, C. W. and Taylor, D. (1984) In *Fatigue Crack Growth Threshold Concepts*, TMS-AIME, USA, p. 433
13. Ohta, A. *et al.* (1982) *International Journal of Fatigue*, **4**, 233
14. Fatigue Data Sheets No. 21 and 31; *Fatigue Crack Properties for Butt Welded Joints*, published 1980, The National Research Institute for Metals, Tokyo, Japan
15. Usami, S. (1985) In *Current Research on Fatigue Cracks*, Society of Materials Research, Japan, p. 101
16. Tsubota, M., King, J. E. and Knott, J. F. (1984) *First Parsons International Turbine Conference* (PIC1), Institute of Mechanical Engineers, London, p. 189
17. Knott, J. F. *Case Studies in Defence and Transport Industries*, p. 279
18. King, J. E. (1987) *Materials Science and Technology*, **3**, p. 750
19. Knott, J. F. (1988) In *Fatigue 87*, EMAS, Warley, UK, p. 497
20. Palmberg, B., Blom, A. F. and Eggwertz, S. (1985) In *Probabilistic Fracture Mechanics and Reliability*, Martinus Nijhoff, Chapter II

Index